# Operational Ratings and Display Energy Certificates

## CIBSE TM47: 2009

CIBSE
Engineering a sustainable
built environment

The Chartered Institution of Building Services Engineers
222 Balham High Road, London SW12 9BS

© July 2009 The Chartered Institution of Building Services Engineers London

Registered charity number 278104

ISBN: 978-1-906846-06-0

Typeset by CIBSE Publications

Printed in Great Britain by The Charlesworth Group, Wakefield, West Yorkshire WF2 9LP


Cover illustration: City Hall, London; photograph by Andreas Stirnberg (www.imagesource.com), sourced through Photolibrary (www.photolibrary.com)

## Note from the publisher

This publication is primarily intended to provide guidance to those responsible for the design, installation, commissioning, operation and maintenance of building services. It is not intended to be exhaustive or definitive and it will be necessary for users of the guidance given to exercise their own professional judgement when deciding whether to abide by or depart from it.

*Printed on recycled paper comprising at least 80% post-consumer waste*

# Foreword

Buildings account for almost half the UK's energy demand, and so the management of energy use in buildings is becoming increasingly important. There is increasing regulatory and market pressure on building owners and occupiers to ensure that the buildings they operate are managed as economically and efficiently as possible.

The EU Energy Performance of Buildings Directive raises the stakes for building energy managers, particularly those operating public buildings. The Directive, which is implemented in England and Wales via the Energy Performance in Buildings Regulations 2007, with parallel provisions covering Scotland and Northern Ireland, requires the production and display of energy certificates for public buildings. In England, Wales and Northern Ireland the Regulations require this Display Energy Certificate (DEC) to be based on metered energy consumption. The resulting certificates reveal how efficiently the building is being operated, and the accompanying advisory report identifies potential improvement measures.

Display Energy Certificates are intended to raise public awareness of energy use in the buildings used by the public sector, and motivate those who operate the buildings to improve the energy performance of the buildings they manage. This comprehensive guidance describes the legal requirements for DECs, the collection of the data and calculation of Operational Ratings and the production of advisory reports. There are also sections which address common questions about DECs, which will be useful to building managers and owners and also to energy assessors. There is also a short section on the Landlord's Energy Statement and Tenant's Energy Review (LES-TER) tool, developed by the British Property Federation in association with CIBSE and others, to enable the owners and operators of multi-tenanted buildings to address the specific energy reporting needs of such buildings.

With the introduction of the Carbon Reduction Commitment in 2010, many of the public bodies who require DECs will also need to provide returns under the CRC. DECs provide a ready-made tool for meeting these aspects of the CRC, and minimising compliance costs.

It is estimated that a fifth of the energy used in European buildings is wasted. At a time of significant economic challenges, there is every incentive to take a greater interest in saving energy, money and carbon emissions in buildings. The Display Energy Certificate and its associated advisory report offers the public sector an opportunity to tackle energy use in the public estate, and this guidance will help them to make the best use of energy certificates to drive economy in the public estate.

Hywel Davies
*CIBSE Technical Director*

# Contributors

Hywel Davies (CIBSE), Lionel Delorme (AECOM), Peter Grigg (formerly BRE),
Colin Lillicrap, Andrew Mercer (AECOM)

# Acknowledgements

Robert Cohen (Camco), Richard Hipkiss (i-Prophets Energy Services), Rob Hunter (Fulcrum Consulting), Steve Irving (AECOM), Ellen Salazar (ES Research and Consultancy), Archie Spence (ABS Consulting) and Nina Wissenden (Enviros Consulting Ltd.)

# Editor

Ken Butcher

# CIBSE Technical Director

Hywel Davies

# CIBSE Director of Information

Jacqueline Balian

# Contents

# Operational Ratings and Display Energy Certificates

## 1 Introduction

The Energy Performance of Buildings (Certificates and Inspections) (England and Wales) Regulations[1–5], and the equivalent regulations in Northern Ireland[6,7], require many public buildings to have Display Energy Certificates (DECs) and advisory reports. This new duty has been introduced as a result of Article 7.3 of the European Directive on the Energy Performance of Buildings[8], commonly referred to as the 'EPBD'.

The regulation on DECs applies to buildings with a total useful floor area over 1000 m² occupied 'by public authorities or by institutions providing public services to a large number of persons and therefore frequently visited by those persons'. The certificate must be permanently displayed in a prominent public position and be renewed every 12 months. It must be accompanied by an advisory report which is valid for 7 years. Both the certificate and the advisory report must be produced by an accredited energy assessor.

This publication provides guidance on the requirements for DECs, along with guidance on their preparation, including what information is needed, who can produce them, what software should be used and what must be done to display them. It is intended to provide assessors with good practice guidance on the whole process of assessing a building for the purposes of producing a DEC in accordance with the approved methods and procedures. It will be of interest to all those involved in the preparation and production of DECs. Scotland has adopted a different approach, with Energy Performance Certificates (EPCs) being used for display purposes instead of DECs, so this guidance is not relevant to Scotland.

The contents of this TM are as follows:

— Section 1 provides a general introduction.

— Section 2 gives an overview of the requirements for DECs and advisory reports, as set out in the Energy Performance of Buildings (Certificates and Inspections) (England and Wales) Regulations (as amended)[1–5]. Those requirements are matched almost exactly by the equivalent Regulations for Northern Ireland[6,7], although the dates of introduction are different. This section reviews the statutory requirements as they relate to DECs, with a brief explanation of how each requirement may be met.

— Section 3 describes the production of the DEC, using the approved software tools that are available for the purpose.

— Section 4 describes the information required to produce a display certificate. It also describes how the Operational Rating can be modified to take account of weather, occupancy patterns and separable energy using activities in the building.

— Section 5 describes the calculation of the Operational Rating of a building. This rating, based on the actual metered energy consumption of the building and its floor area, compares the building to a benchmark, which is based on typical energy use in buildings of that type.

— Section 6 gives guidance on the production of advisory reports and the information which should be included within them.

— Section 7 describes the procedures for assessor accreditation and lodgement of the certificates, which are placed on a national register.

— Section 8 explains how the Landlord's Energy Statement (LES) and Tenant's Energy Review (TER) enable landlords and tenants to identify how energy is being used in multi-tenanted buildings.

— Section 9 provides answers to frequently asked questions concerning all aspects of DECs, ORs and advisory reports.

There are a number of guidance documents available from the Department of Communities and Local Government (CLG) which relate to the implementation of the energy performance in buildings directive. Most relevant to Display Energy Certificates are the introductory guidance *Getting ready for DECs*[9] and the more detailed *Improving the energy efficiency of our buildings: A guide to Display Energy Certificates and advisory reports for public buildings*[10]. At the time of publication, these are available from the 'Planning, building and environment' section of the CLG website (http://www.communities.gov.uk) as free downloads.

This publication provides more extended coverage than the CLG documents, and also provides some guidance on interpretation of the Regulations and the various requirements relating to Display Energy Certificates and assessor accreditation. It also provides some guidance to building owners and managers on the potential benefits of the Display Energy Certificates and advisory reports, and their value to an organisation in identifying opportunities for saving energy, reducing energy bills and cutting carbon emissions.

# 2      Overview of the requirements

This section provides an overview of the requirements for Display Energy Certificates (DECs) and advisory reports as set out in the Energy Performance of Buildings (Certificates and Inspections) (England and Wales) Regulations 2007[1], as subsequently amended[2–5]. It reviews all of the regulatory requirements relating to Display Energy Certificates, and provides a brief explanation of the scope and extent of these requirements.

Unless stated otherwise, all references to 'The Regulations' or 'Regulation X' are to the Energy Performance of Buildings (Certificates and Inspections) (England and Wales) Regulations[1]. Subsequent amendments[2–5] change the dates at which various provisions come into force and alter the scope of the requirements for DECs in the first year of operation.

## 2.1      What are Operational Ratings, Display Energy Certificates and advisory reports?

> Regulation 16 'applies to buildings with a total useful floor area over 1000 m² occupied by public authorities and by institutions providing public services to a large number of persons and therefore frequently visited by those persons.' — Regulation 16(1)

An Operational Rating (OR) is a numeric indicator of the annual carbon dioxide emissions of a building caused by its consumption of energy, compared to a value that is considered typical for that particular type of building. It is based on measured annual energy consumption per unit of floor area of the building.

A Display Energy Certificate (DEC), is a certificate that shows the energy performance of a building, based on the Operational Rating, on a graphical scale from A to G. It also gives the Operational Rating, the Asset Rating if the building has one, historical energy performance data for the preceding two years, if available, and summary information about the energy used by the building.

An advisory report is a report produced to accompany a DEC and sets out recommendations for improving the energy performance of the building.

These are all explained in greater detail below.

### 2.1.1      What does a DEC look like?

> Regulation 17 sets out the mandatory content of a display energy certificate as follows:
>
> (a)    (i) the operational rating and
>
>        (ii) the asset rating (where it exists, see regulation 18) of the building expressed in ways approved by the Secretary of State […];
>
> *(box continues)*

> *(continued)*
>
> (b)    the operational ratings for the building which were expressed in any certificates displayed by the occupier during the two years before the nominated date;
>
> (c)    a reference value such as a current legal standard or benchmark;
>
> (d)    the reference number under which the certificate has been registered in accordance with Regulation 31; this reference number will be related to the property's unique property reference number (UPRN) which is available from the central register.
>
> (e)    the address of the building;
>
> (f)    an estimate of the total useful floor area of the building;
>
> (g)    the name of the Energy Assessor who issued it;
>
> (h)    the name and address of the Energy Assessor's employer, or, if he is self-employed, the name under which he trades and his address.
>
> (i)    the date on which the certificate was issued;
>
> (j)    the nominated date;
>
> (k)    the name of the approved accreditation scheme of which the Energy Assessor is a member.

For an explanation of the nominated date and measurement periods, see section 5.3.3.

Figure 1 shows an example Display Energy Certificate, with a number of key features of the certificate indicated. As well as the Operational Rating and the A–G banding described earlier, the certificate identifies the occupier and building to which it relates (A), the unique reference number (B) and various administrative details including the nominated date (E). It also gives previous operational ratings (D), if available, and a summary of the energy used for heating the building, for other electrical uses, and percentages of energy generated by renewables (F). A histogram (C) shows the carbon dioxide emissions over the previous three reporting periods, indicating the emissions by fuel type, and also indicating whether any on-site renewables have contributed to the energy used by the building.

### 2.1.2      Operational Ratings

> Regulation 15 defines an Operational Rating as 'a numeric indicator of the amount of energy consumed during the occupation of the building over a period of 12 months … calculated according to the methodology approved by the Secretary of State….'

The Operational Rating (OR) is 'a numerical indicator of the actual annual carbon dioxide emissions' from the building. It is based on actual metered energy consumption by the building occupiers over a period of 12 months. This is compared to the performance of a typical building of the same type, the benchmark. A building which uses exactly the same energy as the benchmark building will have an OR of 100. Higher consumption leads to a higher

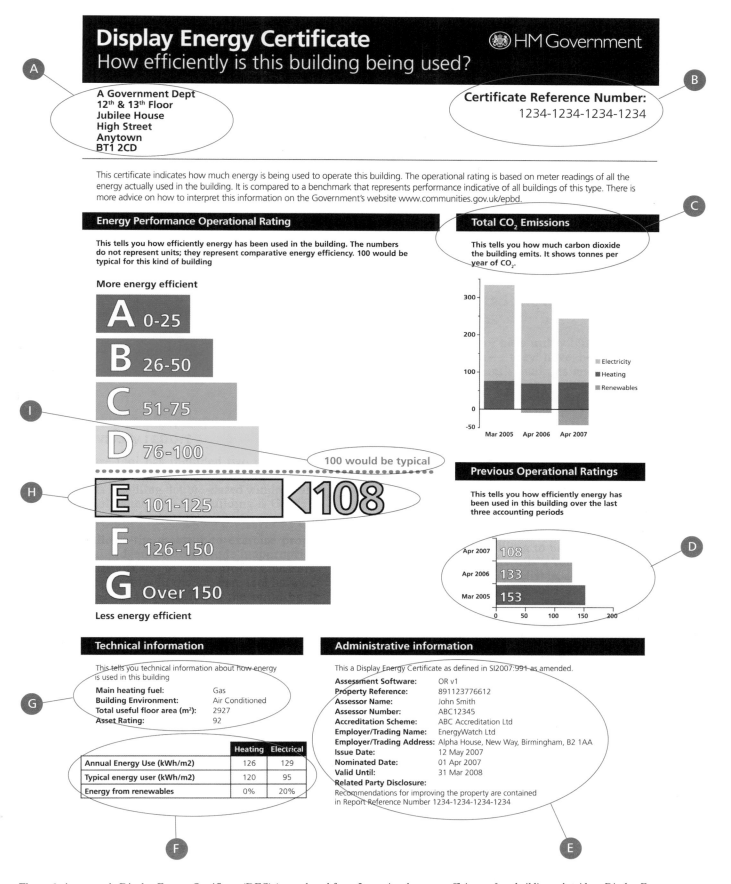

**Figure 1** An example Display Energy Certificate (DEC) (reproduced from *Improving the energy efficiency of our buildings: A guide to Display Energy Certificates and advisory reports for public buildings*[(10)] (Crown copyright 2008))

rating, and lower consumption gives a lower rating. The rating is based on carbon dioxide emissions since this is a key driver for UK energy policy.

The Operational Rating is shown on a scale from A to G, where A indicates the lowest $CO_2$ emissions (best result) and G the highest $CO_2$ emissions (worst result), see item 'H' in Figure 1.

The OR must be calculated according to the methodology approved by the Secretary of State, i.e. by an accredited energy assessor using a software tool for the calculation which has been approved by the Secretary of State. Section 5 of this publication describes the calculation of the Operational Rating in detail, and section 7 describes the procedures for assessor accreditation.

### 2.1.3    Display Energy Certificates

The main feature of the DEC is the familiar A to G energy rating scale (also found on many white goods, such as refrigerators), as described above.

The DEC also shows the building's Asset Rating if one is available (Figure 1, item 'G'). The Asset Rating is a calculation of the intrinsic energy performance of the building based on the performance of its envelope and fixed building services under common operating and weather conditions. For more information on Asset Ratings see *Energy and carbon emissions regulations: A guide to implementation*[11].

The approved methodology also requires that the certificate includes a 'related party disclosure', (see item 'E') to ensure that the public can readily identify any contractual link between the energy assessor and the occupier, beyond the commissioning of the DEC. This covers situations where accredited in-house staff carry out the assessment.

### 2.1.4    Advisory reports

> 'An advisory report is a report issued by an Energy Assessor after his assessment of the building, which contains recommendations for improvement of the energy performance of the building.'

The advisory report accompanies the DEC and includes a list of recommendations to improve the energy performance of the building. The report includes zero and low-cost operational and management measures, possible upgrades to the building fabric or services, and opportunities for low and zero carbon (LZC) technologies. The advisory report is valid for a period of seven years beginning with the date it is issued. The content of the advisory report is considered in detail in section 6.

## 2.2    What is a building?

> '...'building' means a roofed construction having walls, for which energy is used to condition the indoor climate, and a reference to a building includes a reference to a part of a building which has been designed or altered to be used separately.' — Regulation 2

To fall within the scope of the Regulations a 'building' must have a roof and walls, and use energy to condition the internal environment using heating, mechanical ventilation or air conditioning. Provision of lighting or domestic hot water without 'conditioning' does not fall within the scope of this definition.

This definition excludes structures that have permanent openings or that do not use energy to provide heating, cooling or ventilation, for example:

— A multi-storey car park would generally not be classed as a building, since it is naturally ventilated by large permanent openings in the walls (even if it uses mechanical ventilation to improve fresh air provisions).

— A basement car park may be classed as a building, if it has no permanent openings to the outside and uses mechanical ventilation.

— A partially enclosed car park area with lighting but no mechanical ventilation is not classed as a building. Lighting is not considered to 'condition' the indoor climate.

— An unconditioned warehouse or storage facility.

For the purposes of DECs, the definition of a building includes a part of a building which is designed or altered to be used separately, where that part has a total useful floor area in excess of 1000 m². This may be indicated by the accommodation having its own access, by separate provision of heating and ventilation or where there is shared heating and ventilation but with independent occupier control. The part may be considered to be separate even if toilet and kitchen facilities are shared. See section 9.6 for further guidance on the issue of campuses or sites with several buildings such as schools and hospitals.

*Example*

Where a public authority occupies four floors of a 10-storey office building, according to the Regulations 'the building' would comprise all 10 storeys, unless the four floors occupied by the public authority were designed or altered to be used separately, or where the energy to the four floors was sub-metered separately and the actual energy used by the public authority derived from the metered data. However, the Regulations are deemed to cover partly occupied buildings, so that the public authority occupier would generate a DEC and advisory report based on the energy demand for the part of the building it occupies. If the areas occupied in the building are not contiguous horizontally or vertically then the areas and energy use should be aggregated and a DEC produced for the publicly occupied area. Calculation of an OR based on pro rata area is described in section 4.1.5.

Where sub-metering is not available, then the *Landlord's Energy Statement*[12] approach developed by the Usable Buildings Trust in conjunction with the British Property Federation and CIBSE can be used to derive the data, see section 8.

Section 5 of the CLG's *Improving the energy efficiency of our buildings: A guide to Display Energy Certificates and advisory reports for public buildings*[10] gives a number of illustrated examples along with the CLG guidance on how the duty to produce DECs should be discharged in each case.

## 2.3 Which buildings and occupiers require a DEC?

> Regulation 16(1) states that the requirements for a DEC and an advisory report apply 'to buildings with a total useful floor area over 1000 m² occupied by public authorities and by institutions providing public services to a large number of persons and therefore frequently visited by those persons.'

This definition includes two separate categories of buildings:

— those occupied by public authorities

— those occupied by 'institutions providing public services to a large number of persons'.

A 'public authority' is a body whose functions are of a 'public nature' and would include government departments, executive agencies, local authorities, the National Health Service, maintained schools and other educational establishments, and a variety of other public bodies. Guidance given in the context of the Freedom of Information Act[13] (FoI) can be used to identify whether an organisation qualifies as a public authority. Reference should be made to the Act itself and the accompanying guidance, which is available from the Ministry of Justice website (http://www.justice.gov.uk/guidance/guidancefoi.htm).

In summary, the Freedom of Information Act would regard the following as public authorities:

(a) A body set up by the Crown, statute, a government department, the National Assembly for Wales, or a Minister; and where at least one appointment to the organisation is made by the Crown, a government department, the National Assembly for Wales or a Minister. Such organisations are listed in a schedule to the Act that is updated regularly. Note that the FoI guidance excludes from its schedule of public authorities certain elements of the Ministry of Defence.

(b) Publicly-owned companies, i.e. companies either wholly owned by the Crown or by a public authority listed in the schedule referred to in (a) above.

A public service is one provided directly or indirectly by government (central or local) to citizens, fully or partly funded via the public purse. The proportion of public funding provided is irrelevant, and any public funding, however small or however obtained, constitute the service being provided as a public service.

While section 9 gives some further guidance, it should be noted that the ultimate definitions of 'public access' and 'frequently visited' can only be settled by the courts.

The area must be measured as the 'total useful floor area' (TUFA), which is defined as the total area of all enclosed spaces measured to the internal face of the external walls, including sloping areas such as staircases, galleries and auditoria, for which the plan area should be used. This is the same measure of area as used for Building Regulations purposes (see section 5.2 for details), and corresponds to the Royal Institution of Chartered Surveyors' measure of gross internal area.

Implementation of a government programme may be achieved by delegating powers or duties through contractual arrangements. This does not change the public nature of those powers or duties. So a private hospital or clinic admitting NHS patients is providing a public service.

The building must also be 'frequently visited'. Thresholds have not been set to define 'large number of persons' or 'frequently visited'. All buildings where members of the public (i.e. as private individuals rather than as employees or in their trade or professional capacity, e.g. making deliveries or carrying out maintenance) have a right of access to the premises, either to benefit from the public service (e.g. a public swimming pool) or to discuss its provision (e.g. a benefits office), are likely to be considered to be 'frequently visited' by the public.

There is no requirement to produce a DEC for a building which is never, or only exceptionally, visited by the public. CLG has indicated a wish to encourage the provision of the information contained in a DEC, even where it is not legally required, and that if there is doubt about whether a DEC is required, it would be good practice to produce one.

When deciding whether the 'building' (in whole or in separate part) falls within the 1000 m² limit, the assessment of the qualifying area is not limited to that part of

---

### Case study

There are two possible situations for a public authority:

(1) Where the public authority occupies the whole of the building as a tenant and is responsible for all of the energy use and associated payments. In this case, the public sector occupier is responsible for obtaining the DEC for public display and will use the energy data metered in the usual way. If the landlord provides some services and does not charge for this energy use directly then the landlord may need to provide information to the tenant on energy consumption. The landlord has a duty to supply this information under Regulation 50.

(2) Where the public authority only occupies part of the building with other floors occupied by other public or private tenants the position will be more complex. In these situations the occupier still has the responsibility to prepare a DEC using any energy consumption data that has been obtained from local meters, and to ask the landlord to provide a statement indicating the amount of energy that would be allocated to the tenant, e.g. on the basis of floor area. The survey to identify cost-effective improvements for inclusion in the advisory report should not only cover the equipment within the tenant's area. Ideally, the dialogue will be initiated between tenants and landlords on how to reduce the energy consumption for the building as a whole.

the occupied space to which the public has access, but includes any 'back-of-house' areas that support the face-to-face services. For example, in a Post Office, the relevant area is not just the counter and waiting areas, but includes the mail delivery area, sorting office, administration office, maintenance and staff rest areas etc.

If separate areas in the same building are occupied by the same qualifying organisation, then the floor area used to assess whether the building qualifies under the regulation should be based on the combined floor area, i.e. if two parts are each 800 m$^2$ and occupied by the same organisation, then the building qualifies under the regulation, irrespective of whether the separate areas are contiguous or not.

If a qualifying organisation occupies only a part of a physical building that is not designed for separate use, but that part exceeds 1000 m$^2$, then the Regulations apply. The Department for Communities and Local Government has produced guidance which contains a number of examples of part occupation and multiple public tenancies within buildings, with advice on the appropriate approach to take to the provision of display energy certificates. This is contained in chapter 5 of the CLG's *Improving the energy efficiency of our buildings: A guide to Display Energy Certificates and advisory reports for public buildings*[10].

In all those instances where occupiers are unsure whether a DEC is expected to be provided, government recommends that it would be good practice to produce a DEC as under the terms of the EPBD the public sector should set an example.

*Landlord's Energy Statement*

The Usable Buildings Trust, in conjunction with the British Property Federation (BPF) and CIBSE has developed a voluntary code, the 'landlord's energy statement' (LES), that would help landlords fulfil their obligations as landlords. Regulation 50 places a duty to cooperate on those with an interest in or occupation of the building to help facilitate the production of the DEC. The approach set out in the LES offers a suitable procedure for a landlord to provide the necessary information to enable the tenant to produce the DEC and to comply with the requirements of Regulation 50[1]. For further information on this approach see section 8. Details may be found on the LES-TER website (http://www.les-ter.org).

## 2.4    When are DECs required?

> Regulation 16(2) (as amended) states that 'on and after 1 October 2008 every occupier of any building to which this regulation applies must—
>
> (*a*)    have in its possession or control at all times a valid advisory report; and
>
> (*b*)    display at all times a valid display energy certificate in a prominent place clearly visible to the public

This means that occupiers of buildings covered by the Regulation had a duty to obtain and display a DEC and to be in possession of a valid advisory report by 1st October 2008.

Voluntary DECs and advisory reports may be produced and displayed in buildings that do not qualify under the Regulations. Such voluntary DECs and advisory reports must be produced in full compliance with the approved methodology, so only an accredited energy assessor will be able to produce a voluntary DEC and advisory report, and the documents produced will be lodged on the national register.

## 2.5    Who is responsible for obtaining the DEC and advisory report?

> Regulation 16(2) (as amended) states that 'Except where regulation 18(3) applies, on and after 1 October 2008 every occupier of any building to which this regulation applies must—
>
> (*a*)    have in its possession or control at all times a valid advisory report; and
>
> (*b*)    display at all times a valid display energy certificate in a prominent place clearly visible to the public

This places a duty on the occupier to display a DEC (in a prominent place clearly visible to the public) and control, or possess, an advisory report.

*Example:*

(*a*)    Where a private organisation (e.g. a private hospital) provides a public service (e.g. treatment of NHS patients), the responsibility for complying with the Regulations is with the private provider.

(*b*)    Where the building is owned and operated under a Private Finance Initiative (PFI) contract (e.g. a hospital), it is the occupier (the NHS Trust in this example) of the building who must comply with the Regulations, although the PFI provider has a duty under Regulation 50 to assist.

## 2.6    Who should produce the DEC and advisory report and how?

> Regulation 17(1)(d) states 'a display energy certificate must be issued by an Energy Assessor who is accredited to produce display energy certificates for that category of building'; and regulation 19 states that 'an advisory report is a report issued by an Energy Assessor after his assessment of the building …'. Part 5 of the Regulations contains detailed requirements for Energy Assessors. Regulation 25(1) requires that Energy Assessors 'be a member of an accreditation scheme approved by the Secretary of State.'

A current list of approved accreditation schemes is available from the Communities and Local Government (CLG) website (http://www.communities.gov.uk). At the time of writing, CIBSE Certification Ltd. (http://www.cibsecertification.co.uk) operates a scheme for Display Energy Certificates as well as Energy Performance Certificates and air conditioning inspections. Approved accreditation schemes may elect to accredit energy assessors for specific building sectors, for example for healthcare, office buildings, etc.

The skills and qualifications required of accredited energy assessors are based on the National Occupational Standards (NOS) which are available from the CIBSE Certification or CLG websites. Experts who already have the appropriate skills and qualifications may be accredited via the 'approved prior experiential learning' (APEL) route.

Staff employed by the qualifying organisation are allowed to produce the DEC and the advisory report, provided that:

(*a*)      the individual is an energy assessor accredited by an approved accreditation scheme

(*b*)      the DEC includes an adequate 'related party disclosure', whereby the person preparing the DEC indicates any relationship to the person who commissioned the DEC, in the same way as Regulation 26 requires 'related party disclosure' for Energy Performance Certificates, see Figure 1, item 'E'.

The accredited energy assessor must produce the DEC and advisory report according to the methodology approved by the Secretary of State, using approved software. 'ORCalc' is an approved software package that has been developed by government and is freely available to accredited energy assessors from their approved accreditation schemes.

A software specification is available from government for those organisations wishing to develop their own proprietary software, e.g. to embed an Operational Rating software module into an existing monitoring and targeting system (see http://www.ukreg-accreditation.org for details). All proprietary software must be approved by the Secretary of State. Accredited energy assessors wishing to use proprietary software must ensure that the software has been approved by the Secretary of State.

## 2.7      To whom should the DEC be submitted?

When the DEC and the advisory report are complete, the accredited assessor produces (via the accreditation scheme) a data file which is lodged on the national register of certificates and reports operated by Landmark® under contract to the CLG. For more information on the register, go to the *Non-Domestic Energy Performance Certificate Register* website[14] (https://www.ndepcregister.com).

The energy assessor submits the documents generated to his/her accreditation scheme for quality assurance. Once the accreditation scheme is satisfied with the energy assessor's submission, it submits the documents for lodgement on the government's national register. On successful lodgement of the documents, the national register informs the energy assessor and the accreditation scheme that the documents can be issued to the occupier for display. This process is handled automatically by the accreditation scheme and the national register.

## 2.8      Where are DECs to be displayed?

> Regulation 16(2)(b) requires qualifying organisations to 'display at all times a valid display energy certificate in a prominent place clearly visible to the public.'

The DEC should be placed permanently on display in the reception area, preferably on the public side, or in the receptionist's accommodation but where the public can see and read the certificate. The DEC should be no smaller than A3 in size (297 mm wide by 420 mm high).

The occupier may elect to display additional information alongside the DEC, for example to explain that the increase in emissions since the previous year was the result of moving to a double shift operation, but this must be clearly distinguished from the DEC.

Organisations may choose to make their DEC more visible by making them available on their website, see section 9.11 for further information.

## 2.9      The validity of a DEC and advisory report

> Regulation 16(3) states that 'a display energy certificate is valid for a period of 12 months beginning with the nominated date.' Regulation 15 defines the 'nominated date, in relation to a display energy certificate, [as] a date no later than three months after the end of the period over which the operational rating is calculated, which is nominated by the Energy Assessor who issued the certificate.'

This allows three months for data for the chosen 12-month period to be collected and analysed, submitted to the accreditation body and for the display energy certificate to be lodged. Section 5.3.3 considers reporting periods in greater detail. No gap is allowed between successive reporting periods, but an overlap of up to three months is allowed. This may enable the reporting period for the DEC to be aligned with other accounting periods (e.g. for large building portfolios) or with other existing administrative periods.

> Regulation 16(4) states that 'an advisory report is valid for a period of seven years beginning with the date it is issued.'

Unlike DECs, the validity period for advisory reports begins on the date of issue (Regulation 16(4)), which is to be specified on the report, not the 'nominated date' for the DEC. As with DECs, no gap is allowed between successive reporting periods. However, for advisory reports, the overlap between successive reporting periods is not defined, giving qualifying organisations the flexibility to align production of advisory reports with building energy audits. The extended validity of advisory reports allows qualifying organisations to plan for implementing the energy improvement recommendations, and for those recommendations to have sufficient time to influence the operations of the building.

If there is a change of public body occupying the space, they should not display the DEC obtained by the previous organisation. Instead, they should immediately start to collect the necessary data to produce a Display Energy Certificate and advisory report within 15 months of occupation. For the first year of occupation they should produce a DEC that includes the Asset Rating (as new tenants or owners they have bought or rented the building, and so must by law have an EPC with an Asset Rating). However, as they do not have metered energy data they cannot produce an Operational Rating on the first DEC. See sections 2.13 and 9.1 for further information.

A change in the space occupied by a sitting qualifying organisation, for example from occupying two storeys of a 10-storey building to occupying one or three storeys of the 10-storey building, does not invalidate any existing DEC or advisory report, but the subsequent DEC should reflect the change in occupied area.

## 2.10 How can members of the public see a DEC and advisory report?

Under Regulation 34, which sets out the terms for disclosure of documents lodged on the register, members of the public may obtain a copy of a DEC or advisory report, providing they know the unique certificate or report reference number.

'(1) This regulation applies where any person—

(a) requests the keeper of the register to disclose a particular document; and

(b) provides to the keeper of the register the relevant reference number of the document he is seeking to be disclosed.

(2) The keeper of the register may disclose to such a person—

(a) the document the person requested; and

(b) any document of the same kind relating to the same building or part of a building as the requested document, which was registered at any time during the period of 10 years ending on the date of the request.'

Section 2.8 above considers how DECs should be displayed in not less than A3-size so as to be 'clearly visible to the public'. Most organisations will be able to comply with the display requirements with no difficulty by displaying the DEC in the reception area.

All energy certificates, whether EPCs or DECs, are lodged on a national register. The register is not available for general browsing but a copy of an individual certificate may be requested if the unique reference number of the required DEC can be quoted. However, simultaneous public access to several DECs for a specific building is not permitted, although a DEC will include the previous two years ratings where available. Each DEC's unique reference number must be provided to obtain access to that DEC. Members of the public may ask the building occupier to see a copy of the advisory report accompanying the DEC. There is no duty on occupiers under the

Regulations to provide a copy of the advisory report to all members of the public that express an interest. However, members of the public can obtain a copy of the report from the central register if the unique reference number of the advisory report is known. See section 9.12 for examples of public bodies that have made their DECs accessible on the internet.

There is some inconsistency between the objectives of Article 7.3 of the EPBD, to make energy use in the public sector more transparent, and the more restrictive provisions of the Regulations, which constrain direct access to energy certificates.

Occupiers of buildings requiring DECs should be aware that the Freedom of Information Act[13] may be used. It is likely to allow requests for access to multiple DECs and advisory reports, and may also be supplemented by requests for details of the occupier's intentions to implement (or otherwise) the recommendations in the advisory report.

## 2.11 The OR rating scale

There is no legal standard associated with operational performance, and since the available benchmarks are understood in terms of the stock average, the Operational Rating scale is fixed by 2 points: zero emissions at the top of band A and the stock average (i.e. the benchmark for the relevant type of building) at the D/E boundary (Figure 1, item 'I'). Therefore the scale will vary in terms of kW·h per unit of floor area, depending on the type of building being assessed.

For simplicity, CLG has adopted a linear scale. Grade 'G' will have its best limit defined by the linear scale, but will have no worst limit. However, by showing the numeric value of the operational rating, the public can see how poor the building is relative to the 'A' to 'F' grades. It is important to note that in the initial phase of implementation of the Regulations a building that does not have adequate data to produce an Operational Rating will be given a default 'G' grade. It is important to distinguish between these 'default Gs' and G-ratings based on the actual energy use.

Asset ratings for Energy Performance Certificates (EPCs) do not use the same calculation basis as operational ratings for DECs, so no direct correlation can be drawn between Operational and Asset Ratings. However, in both cases a typical building will have a numerical rating of 100, so there is some degree of comparability relative to the wider stock. Where a building with an Asset Rating of 70 achieves an OR of 120 this might well indicate that the building is being operated inefficiently.

## 2.12 What do advisory reports look like?

Regulation 19 defines an advisory report as 'a report issued by an Energy Assessor after his assessment of the building, which contains recommendations for improvement of the energy performance of the building.'

The advisory report should contain a range of recommendations, including cost-effective energy management measures that may be implemented to improve the energy performance of the property. The report should identify for the occupier what may be done to improve, for example, building energy management, building services etc., therefore reducing energy costs, $CO_2$ emissions, and improving the operational rating.

An advisory report includes the following categories of recommendations, based on the general payback period of each recommendation:

— 0 to 3 years payback: e.g. building energy management measures, such as the installation of energy efficient lamps

— 3 to 7 years payback: e.g. upgrading building services, such as replacing obsolete plant

— more than 7 years payback, e.g. low and zero carbon (LZC) technologies.

Each category should include the energy assessor's selection of the most appropriate improvement measures for the specific building. Generally between five and ten measures will be suggested. The advisory report may also include additional comments by the energy assessor, who may recommend further improvement measures; for example, measures recommended by a previous energy audit. Further guidance on generating advisory reports is given in section 6.

The minimum administrative information to be included on the advisory report is shown on Figure 1.

## 2.13 Requirements for new occupiers

Regulation 17(1) requires that 'A display certificate must (a) subject to Regulation 18, express (i) the Operational Rating; and (ii) the Asset Rating of the building in ways approved by the Secretary of State under Regulation 17A of the Building Regulations 2000.'

Regulation 18 sets out the requirements for new occupiers. Regulation 18(1) states that 'Regulation 17(1)(a)(i) does not apply in relation to a Display Energy Certificate which is displayed by an occupier of a building at any time before it has been in occupation of the building for 15 months.'

This means that since January 2009, a new occupier who is required to obtain a DEC has 15 months from first occupation to display a certificate with an Operational Rating. This allows a maximum of three months after the first anniversary of taking occupation to commission the DEC and obtain it and the advisory report.

However, the property will need to have an Energy Performance Certificate for the space to be let. This will include an Asset Rating, which will be accompanied by a recommendations report giving recommended measures for cost effective energy improvements. The Asset Rating should be entered on the first DEC, which will not include an Operational Rating. Within 15 months of occupation, the public authority must display a DEC

based on an Operational Rating calculated from metered energy data, see section 9.1.

## 2.14 Enforcement

Part 7 of the Regulations details enforcement duties, powers and penalties. The following key enforcement principles and mechanisms apply to occupiers required to produce Display Energy Certificates. The enforcement authority in England and Wales is the local authority trading standards department. In Northern Ireland the enforcement authority is the Department of Finance and Personnel, or any person authorised in writing by the Department in a local government district. Non-compliance is a civil offence as detailed in the *Explanatory Note to the Regulations*[1].

Enforcement powers rest with local authorities as defined in Regulation 38(1) which states that 'Every local weights and measures authority is an enforcement authority [...].'

Regulation 39(1) empowers authorised officers of an enforcement authority to require a person who appears to qualify under the display energy certificates requirements (Part 3 of the SI) to produce for inspection a copy of an advisory report. The authorised officer should have access to the DEC for inspection as regulation 16(2)(b) requires that a DEC be displayed 'at all times in a prominent place clearly visible to the public.' Regulation 39(4) gives seven days to a person who appears to qualify under the display energy certificates requirements to comply with the authorised officer's requirements.

Regulation 40(1) states that 'An authorised officer of an enforcement authority may, if he believes that a person has committed a breach of any duty under Regulation [...] 16(2) [...] give a penalty charge notice to that person.' Regulation 43(1) sets out the level of the penalty charge which shall be specified in the penalty notice as follows:

• Regulation 43(1)(b) '... in relation to a breach of a duty under regulation 16(2)(a) [with regard to advisory reports], £1000;'

• Regulation 43(1)(c) '... in relation to a breach of a duty under regulation 16(2)(b) [with regard to DECs], £500;'

The duties imposed on qualifying organisations to comply with the Regulations remain applicable after the organisation has been found guilty of breaching the Regulations and paid any penalty charge. Multiple penalty charge notices may therefore be given until the organisation complies with the Regulations, see section 9.10.

Under Regulation 44, the recipient of a penalty charge notice may give notice to the enforcement authority requesting a review. Regulation 45(1) states that 'If after a review the penalty charge notice is confirmed by the enforcement authority, the recipient may, within the period of 28 days [...] appeal to the county court against the penalty charge notice.'

Public bodies are subject to the provisions of the Freedom of Information Act[13]. Local authorities will need to consider the potential role of elected members in seeking information about compliance. Other public bodies may find themselves receiving requests for information from non-governmental organisations, the press and interested members of the public.

# 3        Producing the DEC

To produce a Display Energy Certificate, an accredited energy assessor must use government approved software to process the information required to produce the Operational Rating and the certificate and to create the advisory report. This section explains the general requirements relating to energy assessors, explains the sources of software and outlines the requirements for the production and lodgement of certificates and advisory reports.

## 3.1        Energy assessors

'An energy assessor must be a member of an accreditation scheme approved by the Secretary of State.' — Regulation 25(1)

In addition, Regulation 26 creates a duty to disclose any personal or business relationship with the person commissioning or procuring the certificate. Regulation 27 creates a duty of care by the assessor when carrying out energy assessments.

Display Energy Certificates and advisory reports may only be produced by energy assessors who are accredited by a government-approved accreditation scheme such as the CIBSE Low Carbon Energy Assessors Scheme. Building occupiers must ensure they employ an energy assessor accredited specifically to produce DECs and advisory reports as energy assessors may be accredited to produce other documents such as energy performance certificates and recommendation reports, or to undertake air conditioning inspections. Energy assessors should only undertake work for which they are accredited. An energy assessor who is not appropriately accredited will be unable to lodge the DEC on the national register. Assessors have a duty to ensure that they are working within their own competence at all times.

Energy assessors are accredited by one of the government-approved accreditation schemes. An up-to-date list of approved accreditation schemes is available from the CLG website (http://www.communities.gov.uk). CIBSE Certification Ltd. is accredited to approve energy assessors for EPCs, DECs and air conditioning inspections. CIBSE accredited energy assessors are known as CIBSE Low Carbon Energy Assessors. For more information visit the CIBSE Certification Ltd. website (http://www.cibse certification.co.uk). This gives details of CIBSE Low Carbon Energy Assessors, and may be used by a building occupier to find a suitably accredited energy assessor to produce a Display Energy Certificate or an Energy Performance Certificate, as well as to find an air conditioning inspector.

## 3.2        Software for calculating Operational Ratings and producing Display Energy Certificates

Energy assessors must use government-approved software to calculate ORs, produce DECs, and to prepare advisory reports. Government has commissioned the development of 'ORCalc' for this purpose, and a number of other organisations have produced their own proprietary software for this task. To be used for the production of DECs and/or advisory reports, proprietary software packages must be approved by the Secretary of State.

### 3.2.1        Government's software

Government has produced a free-issue software package called 'ORCalc', for use by suitably accredited energy assessors to generate display energy certificates, advisory reports, and data files (in XML format) for lodgement on the national register. This software is available to accredited energy assessors through their approved accreditation schemes.

The software requires a Central Information Point (CIP) file and a 'plug-in key' to generate DECs and advisory reports. The CIP file contains reference data for use in the calculation of the Operational Rating, including benchmarks and monthly degree-day data. The CIP file must be updated monthly to include the latest available degree-day data. This can be done by accredited energy assessors downloading the CIP file from the national register's website (http://www.ndepcregister.com) and saving the updated file every month.

The government software checks that suitable degree-day data are available from the CIP file up to the end date of the specified assessment period of the Operational Rating. The software does not allow completion of the OR calculations, DEC and advisory report generation if up-to-date degree-day data, to the specified end date of the assessment period, is not available. This is one reason for the three months permitted between the nominated date and the date on which the DEC must be displayed.

In such cases the energy assessor must download the latest CIP file, alter the end date of the assessment period, or wait for the up-to-date degree-day data to be made available.

Accreditation schemes provide a software 'key' to their accredited energy assessors to enable them to produce complete Display Energy Certificates and advisory reports, which are then lodged in the national register as XML files by the accreditation scheme's software. The Operational Rating software may be used without the key, but it then produces DECs and advisory reports marked 'draft', which are not acceptable for public display or for lodgement. Users should note that as with SBEM for Energy Performance Certificates (EPCs), the accreditation schemes have to obtain the software key from the government's software developers, and do not generate this key themselves.

### 3.2.2 Third party software

Government has produced a software specification detailing the procedures which the software is required to implement to calculate the Operational Rating, and produce Display Energy Certificates and advisory reports. The specification is intended for those organisations wishing to develop their own proprietary software; for example, to embed an Operational Rating software module into an existing monitoring and targeting system. It allows interested software houses to produce third party software packages for the production of Operational Ratings, DECs and advisory reports.

All third party software must be approved by the Secretary of State. Accredited energy assessors wishing to use such software must ensure that the particular software release has been approved by the Secretary of State. For further information about the specification, the approval process and requirements for third-party software for DECs and EPCs, as well as full details of the software which has been approved visit the Building Energy Calculation Software Approval Scheme website (http://www.ukreg-accreditation. org/Index.html).

Software houses interested in the development of proprietary software to generate Display Energy Certificates and/or advisory reports will need access to the CIP file to ensure their software can interact with the CIP. Enquiries about software validation and approval (only) should be directed to: inform@UKReg-accreditation.org

Software houses will also need to ensure that their software can interact with the accreditation schemes' key to generate and lodge complete Display Energy Certificates and advisory reports, and should liaise with accreditation schemes directly.

## 3.3 Lodging a DEC: legal requirements

Once a complete DEC and/or advisory report and their respective accompanying data files have been produced by the energy assessor, they must be submitted to the accreditation scheme. This submission process is left to the accreditation scheme's discretion and therefore may vary. For information on CIBSE Certification Ltd. procedures, CIBSE accredited energy assessors should contact CIBSE Certification Ltd (http://www.cibse certification.co.uk).

The accreditation scheme then submits the DEC or advisory report data files to the central register for lodgement operated by Landmark®. In parallel, the accreditation scheme is required to undertake initial quality assurance checks on the documents produced by the energy assessor. CIBSE Certification Ltd. runs a number of checks on each of the documents produced to identify possible errors, and certificates and reports that contain unusual or unexpected information.

Once lodged on the central register the documents are available for retrieval by members of the public that have obtained the document's unique reference number. The availability of the documents on the central register also provides a check for building occupiers to ensure that the energy assessor has completed his/her duties.

Once certificates or reports have been lodged then they may be made available to the building occupier for display or keeping.

## 4 Gathering data and evidence

In order to calculate the Operational Rating and produce a DEC and advisory report, the energy assessor needs to gather various data about the building, including its size, use, occupancy and energy consumption. The various data required are described below, along with some guidance on the collection of the data and its use in the calculation of the Operational Rating.

### 4.1 Data required

The energy assessor needs to gather data and supporting evidence under the following headings:

— building category (see 4.1.1)
— location (postcode, building name, address) (see 4.1.2)
— unique property reference number (UPRN) (see 4.1.3)
— energy consumption (meter readings or suppliers estimates) and measurement period (see 4.1.4)
— building (or site) area and how it has been defined (see 4.1.5)
— separable energy uses if any (see 4.1.6)
— recorded hours of occupancy (see 4.1.7).

#### 4.1.1 Building category

The building needs to be placed in one of 29 benchmarking categories. A brief description of the categories is provided within the government's OR calculation software and more details are available, with additional guidance, in CIBSE TM46: *Building energy benchmarks*[15]. This describes each category together with the types of buildings that fall within each category.

#### 4.1.2 Location

The postcode is required to allow the Operational Rating calculation software to adjust the benchmarks for local weather effects.

#### 4.1.3 Unique property reference number

The unique property reference number (UPRN) is required to define the space being assessed, e.g. a part of a specific building that is let separately, and is obtained from the government's national register (http://www.ndepcregister. com). The report's reference number allocated to the DEC or advisory report is also used to check and retrieve previous certificates, Asset Ratings or Operational Ratings from preceding years, as these may need to be included in the DEC.

### 4.1.4 Recording energy consumption and the period of measurement

The energy assessor will need to gather data on the 'annual' consumption of each fuel, whether electric, fossil, bio-fuel or district heating, and also obtain details of the measurement period for each. This information will need to be gathered from historic meter readings or utility bills. These data may need adjusting, according to the provisions of section 5.3.3.

Article 13(3) of the Energy Services Directive[16] requires information to be provided with energy bills, where appropriate, to include historic data (i.e. comparison of a customer's usage with the same period of the previous year) and benchmarking information (i.e. comparison with an average energy user in the same user category) information. The energy assessor may therefore be able readily to obtain the historic data required from utility bills, once the Energy Services Directive has been implemented in the UK.

After the first renewal of a DEC and in subsequent years, the certificate must display relevant ORs for the previous two years, as long as the building has not undergone a change of use or occupier during that time, in which case the previous ORs are no longer relevant or required. These data can be accessed as described in section 4.1.3.

In principle, an accurate meter reading of fuel and electricity should be taken once a year to enable annual energy use to be obtained. For those sites which have half-hourly metering, this will not be a problem. If the meter reading, or other means of measurement, has to be a few days sooner or later than the selected date, but within permitted tolerances, adjustments to give the consumption over a 365-day period are made automatically by the software. The energy assessor will need to record the dates the readings were taken and enter them into the software. The tolerances allowed on the measurement dates are detailed in section 5.3.3 below. It is probably realistic to expect that the tolerances may be tightened over time. Recording energy used by separable energy uses is covered in section 4.1.6.

If heating or cooling is supplied from district heating or cooling installations then the annual readings of the heat meters should be used. These will have to be supplemented by a statement from the system operator of the carbon burden of the energy supplies, i.e. kg of carbon per kW·h of delivered energy. Similarly, if bulk fuels such as oil, LPG, coal or biomass are used, consumption will have to be estimated from stock measurements, or delivery notes.

#### 4.1.4.1 District heating and cooling

Where heating or cooling energy is supplied from district heating or district cooling installations, the suppliers of these services are required to calculate, from their own energy records, a $CO_2$ content per kW·h of energy supplied. Calculations should take account of the annual average performance of the whole system (including all heating/cooling/power generating plant, any heat recovery or heat rejection/dumping and the distribution circuits). The assessment of $CO_2$ content per kW·h should be accompanied by a report signed by a suitably qualified person, detailing how the emission factors have been derived.

The energy assessor will need a copy of this report together with the start and end dates for the measurement period and the kW·h of energy delivered.

#### 4.1.4.2 Non-metered energy consumptions

Solid fuels and in some cases oil will not be metered. The energy assessor will need to obtain records of deliveries and a statement of the stock level at the start and end of the measurement period. The energy assessor will need to obtain a signed statement by a responsible person that the stock level was measured and details of the method used. The same tolerance of $365 \pm 31$ days applies to the measurement period for solid or liquid fuels, see section 5.3.3. The energy assessor will convert the fuel consumption in kg or litres to kW·h and enter this figure together with the start and end dates of the measurement period.

If actual meter readings or utilities suppliers' estimates are not available, the CLG methodology requires a default value to be used. As this is twice the value for a typical building, the result will automatically be a 'G' rating. This has been set so as to provide an incentive for the regular reading of meters in public buildings, which is a fundamental requirement of good energy management practice and public accountability. There should not be any reason to use estimated readings after the first year, as the requirement for meter readings should by then be fully understood.

If the only data available to the energy assessor are outside of the tolerance period, the software will also apply the default values, i.e. an Operational Rating of 200 and a 'G' rating.

### 4.1.5 Building floor area

The Operational Rating is based on the total useful floor area (TUFA) as described in section 5.2.

Other definitions of floor area may be used to normalise energy demand, e.g. net internal floor area used (or occupied as working space) or treated floor area (i.e. gross internal area less unheated spaces). These alternative floor areas should *not* be used as the basis to calculate the Operational Rating. However, such metrics can be used as a way of determining the TUFA where so approved in the methodology. For example, for rented offices, net lettable area is the norm and this can be used in conjunction with a conservative ratio of net to gross area to deliver the required value of the TUFA. In this context, 'conservative' means the maximum likely value of the ratio of the net to gross areas, thereby delivering the smallest TUFA (and therefore a pessimistic operational rating). The allowed ratios are listed in Table 1.

The energy assessor should enter which measure of area is being used into the software and calculate the area from plans or by measurement. Where no plans exist the energy assessor should produce a sketch of the building outline and mark all relevant dimensions on it. Where external dimensions are used it will be necessary to allow for the wall thickness when calculating the gross internal area or one of the allowed alternatives. All calculations should be

**Table 1** Conversion factors for other measures of floor area to TUFA

| Category | Name | Brief description | Approved alternate floor area | Default multiplier applied to obtain TUFA |
|---|---|---|---|---|
| C1 | General office | General office and commercial working areas | Net lettable area* (NLA) as defined in RICS *Code for Measuring Practice*[17] | 1.25 |
| C3 | General retail | General retail and services | Gross floor area measured as sales floor area (SFA) | 1.80 |
| C4 | Large non-food shop | Retail warehouse or other large non-food store | Gross floor area measured as sales floor area (SFA) | 1.80 |
| C5 | Small food store | Small food store | Gross floor area measured as sales floor area (SFA) | 1.35 |
| C6 | Large food store | Supermarket or other large food store | Gross floor area measured as sales floor area (SFA) | 2.00 |

* Also called 'net internal area' (NIA), which is defined in RICS *Code for Measuring Practice*[17] as 'the useable area within a building measured to the internal face of the perimeter walls at each floor level', where 'an area is useable if it can be used for any sensible purpose in connection with the purposes for which the premises are to be used'.

shown on the plan or sketch. The overall Operational Rating is highly sensitive to the floor area measurement, so the energy assessor should take care when calculating the floor area figures, and should ensure that they have the evidence to substantiate the figures used in the calculation.

Where the DEC is only required for part of the building, then the area of that part covered by the DEC is required. Where the building is partly occupied by the public body and some (or all) of the energy is supplied by the landlord to the whole of the building, or to other parts of the building in addition to that part or parts occupied by the public body, then the total floor area of the whole building or of the larger part or parts will be required in order to apportion the energy consumption on a pro rata basis to the public body.

### 4.1.6 Separable energy uses

Separating out certain energy uses is an optional part of the approved Operational Rating methodology which can increase the relevance of the Operational Rating if a building houses certain specified 'process' energy uses which can not currently be meaningfully included in the building's benchmark category.

Separable energy uses are energy uses within a building's overall metered energy consumption which, along with their associated floor area, the Operational Rating methodology allows to be reported separately from the Operational Rating of the building.

This part of the methodology is simply omitted if the building has no metered separable energy use, and all the building's energy is then counted in the assessment of the operational rating. If there are unmetered separable energy uses, this provides an incentive to meter them for future assessments.

The allowed separable energy uses are:

— regional server room

— trading floor

— bakery oven

— sports flood lighting

— furnace, heat treatment or forming process

— blast chilling or freezing.

To be allowed, the energy use must:

— comply with the criteria defining the separable energy use

— have permanently metered energy use

— have meter readings and analysis for the rating period as for the main use

— have associated floor area measured and recorded

— have a collated Separable Energy Record summarising all the above and signed by the occupant's property manager.

No other energy uses may be separated from a building's operational rating assessment. The Operational Rating software requests the energy consumption (by supply type) and floor area of any separable energy use. Energy assessors may only enter this information if they have completed a Separable Energy Record which is signed by the occupant's premises manager.

The energy assessor should also confirm that the efficiency of the separable energy use has been assessed in the past two years, otherwise such an assessment should be recommended in the advisory report.

In cases where the separable energy use is intermittent or seasonal the energy assessor will need to give special consideration to how the measurement period is entered into the software. If the dates of the meter readings or other measures do not span a full year, because for instance the separable energy use is seasonal or summer only, such that outside of the measurement period there is no consumption, it will be appropriate to enter dates which span 365 days. This will avoid the software scaling-up the consumption for the periods when the separable energy use is not in operation, or the software disallowing the separable energy use if its measurement period is outside the permitted tolerances.

As the separable energy consumption is deducted from the total consumption for the building, so the area of the building occupied by the separable use must be deducted from the area measurement for the building. The energy assessor will need to obtain this area by measurement or from plans. Where the separable use is outside, such as sports flood lighting, the area to be deducted will be zero.

### 4.1.7 Occupancy adjustment

Occupancy adjustment is an optional part of the approved Operational Rating methodology which can increase the relevance of an Operational Rating in buildings for which the occupancy differs from the benchmark occupancy value. Where there is robust documentary evidence of the actual occupancy of the building, based on attendance records, survey results or published opening hours, the Operational Rating may be modified by adjusting the listed energy consumption benchmarks according to the actual occupancy.

The energy assessor must obtain attendance records, survey results or published opening hours and calculate the annual occupancy hours. This information is to be collated into an annual occupancy hours record and signed by the occupants' premises manager before the energy assessor uses the occupancy data in the Operational Rating procedure.

To obtain the annual occupancy hours the energy assessor must use the appropriate occupancy measurement systems allocated to each benchmark category, as follows:

(a)   Annual occupancy hours are defined as the number of hours per year that the number of recorded occupants exceeds 25% of the nominal maximum occupancy.

(b)   Annual occupancy hours are defined as the number of hours per year that the premises are fully open to the public according to published opening hours.

If suitably documented evidence of the actual occupancy of the building is not available, or occupancy is the same as the benchmark occupancy, the occupancy adjustment is not applied and the listed (unadjusted) energy benchmarks are used. Where occupancy is higher than the benchmark figure, this provides an incentive to obtain the occupancy data for future assessments.

The Operational Rating methodology uses the following data for each benchmark category:

—   listed energy consumption benchmarks

—   reference occupancy (for the listed benchmarks)

—   limiting (maximum) occupancy

—   limiting (maximum) percentage increase in energy consumption at limiting occupancy.

The benchmark is then adjusted as follows:

—   If the building occupancy is less than or equal to the reference occupancy in the benchmark, then the benchmark figure is used with no adjustment.

—   If the building occupancy is equal to or higher than the limiting occupancy in the benchmark

data then the benchmark is adjusted by applying the limiting percentage increase.

—   For occupancy values in between these two extremes the percentage increase is pro-rated to obtain the adjusted benchmark.

For those categories of buildings where occupancy adjustments are permitted, a maximum occupancy limit is defined. Adjustments for occupation beyond the maximum specified are not possible, as there is a limit to the marginal effect of extended operation on energy use.

Where different parts of the building falling into the same benchmark category have different occupancies the lowest occupancy must be used, unless an assessment of occupancy in each part is made and the occupancies combined using the percentages of overall floor areas, i.e. using an area-weighted average.

For occupancy adjustment of a multi-use building assessment (employing more than one benchmark category), the annual occupancy hours must be calculated as above for each category for which an occupancy adjustment is relevant.

At present it is not possible to take account of the density of occupation in calculating a DEC. This means that intensively used buildings may achieve a higher Operational Rating than an identical, but less intensively used building. The Department for Communities and Local Government is aware of the need to address this in future developments of the benchmarks for DECs.

## 4.2 Application to groups of buildings on a site

The Regulations apply to buildings (or parts thereof designed or adapted to be used separately) and the current legal advice is that, with the exception of the transitional arrangements, there is no scope for applying the approved operational rating methodology to groups of buildings on a site (e.g. a campus or hospital). Each qualifying building must display its own DEC. However, many public authorities have campus-style facilities where, currently, metering is at the site level rather than at the individual building level. In such cases it is reasonable for the Display Energy Certificates to be based on the metered site energy demands, but with the consumption disaggregated into the demands for each building on an area-weighted basis. Over time, metering provision should improve as Part L of the Building Regulations requires additional metering to be installed in both new buildings and existing buildings that are being refurbished. This will allow DECs that are specific to each individual building to be produced.

On some sites, individual buildings may have dedicated metering, and in such a situation, building specific DECs should be produced wherever possible based on the data from these meters, i.e. where:

(a)   the individual building(s) have dedicated metering for all fuels used in the building; *and*

(b)   there is an appropriate benchmark for that building activity; *and* (*c*)     the individual building(s) fall(s) within the public display requirement, unless a voluntary DEC is being produced.

For the remaining buildings on the site, DECs should be based on the whole site consumption, excluding those buildings that are separately assessed. The benchmark should be based on the site benchmark, but adjusted by the emissions associated with the benchmarks for the buildings that have been separately assessed.

For example, if an operating theatre block at a teaching hospital had its own gas and electricity metering, it would be impractical to produce a DEC for the theatre block since benchmark data for theatre blocks are not available, whereas a benchmark for the site, i.e. a teaching hospital, is available. However, if the separately metered building were an administration building (i.e. essentially an office), it could be benchmarked independently. The revised site benchmark for the remaining buildings would then be:

$$\text{BM}_r = \frac{(A_{\text{site}}\,\text{BM}_{\text{site}} - A_{\text{office}}\,\text{BM}_{\text{office}})}{(A_{\text{site}} - A_{\text{office}})} \qquad (4.1)$$

where $\text{BM}_r$ is the revised site benchmark for remaining buildings ($\text{kW·h/m}^2$), $A_{\text{site}}$ is the floor area of the whole site ($\text{m}^2$), $\text{BM}_{\text{site}}$ is the benchmark for the whole site ($\text{kW·h/m}^2$), $A_{\text{office}}$ is the floor area of the administration building ($\text{m}^2$) and $\text{BM}_{\text{office}}$ is the benchmark for the administration building ($\text{kW·h/m}^2$).

When the DEC is updated each year, the procedure described above should be repeated rather than continuing with the status quo. This means that in the above example, if benchmark data had become available for operating theatres in the period since the previous DEC was produced, the theatre block would henceforth be treated separately from the rest of the site

In the context of a site, the advisory report should encourage the preparation of a specific energy survey for each qualifying building as soon as possible, but certainly before the next advisory report is due. If such reports are already available, then the free form entry field in the 'advisory report' software can be used to log the recommendations that have already been made.

*Note*: the requirement for a DEC applies only to buildings over 1000 $\text{m}^2$. Consequently, if the total area of the buildings on a site exceeds 1000 $\text{m}^2$, but this comprises several individual buildings, none of which is on their own of greater area than 1000 $\text{m}^2$, there is no requirement for a DEC to be produced. If one or more buildings exceeds 1000 $\text{m}^2$, a DEC is required for each such building (but not for those with floors area less than 1000 $\text{m}^2$).

# 5     Calculating the Operational Rating

The Display Energy Certificate (DEC) (see Figure 1) is based on the calculation of the Operational Rating (OR) of a building. This section describes the methodology as it is applied by the approved software tools.

Regulation 15 defines the Operational Rating as 'a numeric indicator of the amount of energy consumed during the occupation of the building over a period of 12 months (unless Regulation 18(4) applies) ending no earlier than three months before the nominated date, calculated according to the methodology approved by the Secretary of State for the purposes of Regulation 17A of the Building Regulations 2000'.

It is intended that the energy consumed be based on actual meter readings. The Operational Rating is derived by comparing the energy consumption of the building with the benchmark energy consumption of other buildings representative of its type. In its very simplest form the OR would be expressed as the total annual energy used by the building divided by the area of the building, compared to the energy use per unit area of a building typical of its type.

Buildings often use more than one form of energy. Simply adding together annual consumptions of gas and electricity is unhelpful in any accounting, as the fuels represent different primary energies and costs, and, crucially, give rise to different carbon emissions. To allow different forms of energy to be added together and compared on a common basis, the government has decided that the common unit should be carbon dioxide ($CO_2$) emissions, since this is the most important driver for energy policy. A cost-based rating would be unsuitable because energy supply prices for non-domestic consumers are highly variable and so meaningful comparisons based on cost would be difficult to achieve. The OR is then a measure of the annual $CO_2$ emission per unit of area of the building caused by its consumption of energy, compared to a value that would be considered typical for the particular type of building, i.e:

$$\text{OR} = (\text{building } CO_2 \text{ emissions} / \text{building area}) \times (100 / \text{typical } CO_2 \text{ emissions per unit area})$$

A typical building will therefore achieve an Operational Rating of 100, which places it on the boundary between Grades D and E. A building that resulted in zero $CO_2$ emissions would have an OR of zero (Grade A), and a building that resulted in twice the typical $CO_2$ emissions would have an OR of 200 (Grade G). If the building is a net energy generator, it would still be given an Operational Rating of zero — it is not possible to achieve a rating less than zero. However, the contribution of the renewable sources is noted in the histogram on the DEC showing total carbon dioxide emissions (see Figure 1 item 'C').

Within this simple description, though, each of the terms used requires further clarification so that buildings can be compared on an equitable basis.

## 5.1     Defining the building

The Regulations apply to buildings, or parts of buildings designed, or altered, to be used separately. In an ideal situation each building would have its own energy meters or, where only part of a building is occupied by the authority that needs to display a DEC, that part would be metered separately. Where there are a group of buildings on a site that is metered only at site level, then each building should normally be assessed individually, and the energy used by each building would be determined from the site energy consumption on a simple area-weighted basis. The process of disaggregating the energy on an area-weighted basis means that the OR for each building will be the same and equivalent to the value that would have been obtained if a site-based calculation had been carried out.

In some specific cases, and for a transitional period only, groups of buildings on a site may be assessed on a whole-site basis. In these cases, the use of the term 'building' should be interpreted as referring to the 'site' for the purposes of the calculations and corrections described in the following paragraphs. The term 'building' should also be interpreted as referring to subdivisions of a building where an OR is to be calculated and a DEC for the whole site produced.

## 5.2        Building area

The building area measurement specified in the legislation is the total useful floor area (TUFA). This is the same as the gross internal area (GIA) in common use in commercial property surveying, and for which measurement conventions are based on the RICS *Code of Measuring Practice*[17] and defined in the Building Regulations[18].

Building Regulations Approved Document L2A[19] defines TUFA as the:

> total area of all enclosed spaces measured to the internal face of the external walls, that is to say it is the gross floor area as measured in accordance with the guidance issued to surveyors by the RICS. In this convention:
>
> a.   the area of sloping surfaces such as staircases, galleries, raked auditoria, and tiered terraces should be taken as their area on plan; and
>
> b.   areas that are not enclosed such as open floors, covered ways and balconies are excluded.

Some building sectors commonly use alternative measures of area, notably net lettable area (NLA) for the commercial office sector, and sales floor area (SFA) for retail premises. Where these are the only measurements available for these building types, then the calculation may use standard, conservative, conversion factors to obtain GIA from NLA or SFA. These conversion factors, and the building categories for which they may be applied, are described more fully in section 4.1.5 above. The only alternative to using these defined conversion factors would be to measure and provide the GIA directly.

Alternative normalising metrics, such as numbers of hotel beds, could be used as part of a non-regulatory, voluntary, sector-specific, initiative. However these cannot be used on a DEC nor can they be displayed in a way that might cause any confusion with a statutory DEC.

### 5.2.1        Unconditioned areas

Within the total useful floor area, some covered areas may be untreated (i.e. neither heated, cooled nor mechanically ventilated), and are termed 'accessible unconditioned areas' (e.g. attics and basements). Although the calculation of the OR is not adjusted to take account of these areas

(and they do not appear on the DEC), these areas should be recorded as part of the data entered into the calculation procedure. Each accessible unconditioned area should be recorded together with a description of the purpose of the area, so that these can be included in the output data file and be available for subsequent analysis.

## 5.3        Energy consumption and carbon emissions

The ultimate goal is that all energy flows into (and out of) the building will be metered. However, at least initially, energy consumption estimates provided by the utility companies will need to be used where actual meter readings are not available. During the initial implementation period, estimates may also be needed where the only readings available do not satisfy the requirements for calculating the OR, in particular where they do not satisfy the rules for the measurement period, which are discussed further below.

So that calculations based on estimates can be distinguished from those based on actual measurements, throughout the calculation procedures and in the storage of results, energy consumptions based on actual measurements are 'flagged' with the letter 'A', and those that are estimates provided by utility companies are flagged 'E'. These flags are not shown on the DEC, although the DEC will show that it has been based on estimated energy consumptions where any of the energy consumptions are estimated. It is important to note that estimates must be supplied by the relevant utility company, and not the occupying organisation.

### 5.3.1        The assessment period

The assessment period is the one-year period over which the energies used in the calculation of the OR are considered to have been used, and is aligned with the measurement period of the main heating fuel used in the building. If the main heating fuel is measured over a period of exactly one year, then the assessment period is identical to that measurement period. Otherwise either the start date of the assessment period is taken as the start date of the measurement period of the main heating fuel, or the end date taken as the end date of that measurement period.

### 5.3.2        Separable energy uses

The aim of the OR is to compare the annual energy consumption of the building with that of a building typical of its type. In some cases, though, the building may include activities that consume energy and which are not considered typical of that building type. Including these activities could reduce the validity of the comparison, and so it may be reasonable to subtract these separable energy uses.

In order to be able to isolate and remove the annual separable energy consumption from the total, any separable energy uses must be separately metered. This is to ensure that the adjustment is based on robust evidence and will also encourage the installation of sub-meters. The separable energy measurement period must be within the range $365 \pm 7$ days, and the measurement period must be aligned to begin within ±31 days of the beginning, or end within ±31 days of the end of the assessment period or the separable energy use cannot be accommodated and discounted in the calculation.

The energy demand of measured separable energy uses will not be recorded on the DEC, but will be included in the calculations and in the more comprehensive 'technical table' produced by the software. This will show both the measured demand and the area occupied by separable energy uses that have been deducted from the building totals before calculating the OR. Where there is more than one separable energy use, they should each be individually reported in the calculation data file, so that these might be used later to develop or refine benchmarks.

The facility to deduct a separable energy use and its area from the OR assessment is an option available to the energy assessor. However, to prevent claiming unjustified separable energy uses, the separable energy use must be one that is included on a list of allowable adjustments for the particular building type, included as part of the benchmark information described later. No other energy uses may be separated for the assessment. This ensures a common and consistent approach to the delivery of DECs.

Alternatively, the energy assessor might deal with the space separately, using the same method as for the remainder of the building, using a 'composite benchmark' approach as described in section 5.4.3.

In order for an energy use to be separated from the total energy consumption, the energy use must:

— be one of the separable energy uses defined in the appropriate table of the benchmark information

— comply with the criteria indicated in the benchmark information defining the separable energy use

— have permanently metered energy use

— have meter readings and analysis for the assessment period

— have the associated floor area measured and recorded

— have a collated Separable Energy Record summarising all the above and signed by the occupant's property or premises manager

The Operational Rating procedure requests the energy consumption (by fuel type) and the floor area of any separable energy use. Energy assessors have the option to enter this information only if they have completed a Separable Energy Record, which is signed by the occupant's premises manager, for each. In addition, separable energy uses tend to be energy intensive, and it is therefore highly advisable for the occupier to obtain a thorough survey of the energy use and efficiency with improvement proposals.

### 5.3.3    Energy measurement periods

Calculation of the OR is based on annual energy consumption, which is defined as the energy consumed over the assessment period of 365 days. This constant number of days has been selected in preference to one calendar year, remaining as 365 days rather than 366 in leap years. The ideal situation would be where all energies are metered over the same one-year period. However it is recognised that, at least initially, the different forms of energy consumed are likely to have been measured over different periods, which may be displaced in time from each other. Provided the measurements are within reasonable limits, the calculation accommodates these by extrapolating or interpolating from shorter or longer measurement periods. Displacements in time (or lack of synchronisation) between the measurement periods of different fuels could, however, make the measurements incompatible and so displacement beyond certain limits cannot be used to produce a reliable result. In such cases, and in the initial period, the energy assessor will need to obtain estimates of consumption for a suitable measurement period.

Where energy consumption cannot be measured, or utility company estimates provided, to cover all the significant energy consumed over periods that are suitable and closely aligned with the assessment period, then the DEC will show a default OR and other comparative details equivalent to those of a building with twice the energy consumption of one considered typical for its type. This will, by definition, equate to an OR of 200, and an automatic worst 'G' rating.

The remainder of this section sets out the rules for measurement periods in some detail; those who do need to understand the details of this procedure may wish to proceed directly to section 5.3.4.

The method of extrapolating or interpolating energy use from measurement periods that are not exactly one year depends on the use of the energy. The main heating energy needs to be treated differently to the energies used for other purposes. This is because uses of energy other than for space heating are considered to be relatively constant throughout the year, and so correction can be applied on a pro-rata basis according to the length of the measurement period. The main heating energy is considered to be at least partially weather dependent, and so the measured energy should be corrected in proportion to the number of heating degree days in the measured and in the 'extrapolation' or 'interpolation' periods.

Where there are metered Separable Energy Uses, any metered 'separable' use of the main heating fuel must be subtracted from the total before the degree-day correction is applied. The metered separable energy use must first be adjusted to a measurement period equal to that of the main heating fuel, and this adjustment is applied on a simple pro-rata daily basis, providing energy consumption data is within the allowed tolerances. The degree-day corrected component of the main heating fuel consumption and the 'separable' use of the main heating fuel, itself adjusted to a 365-day period, are then added together to form the adjusted 365-day total consumption of the main heating fuel.

Where the main heating fuel is electricity, and the heating electricity is not separately metered, then a 'default' value for the proportion of the electrical consumption deemed to be used for heating is provided in the benchmark information. This proportion is then treated as being weather dependent, and the degree-day correction applied to this proportion.

For the main heating fuel only, if the measurement period does not start at the beginning of a month, then the number of degree-days in the part-months are obtained in simple proportion to the number of days that are included from those months. So, if the measurement period begins on the 6th day of a month having 31 days, and ends on the 5th day of a month having 30 days, then the number of degree-days in the month during which the measurement period started would be would be multiplied by $[(31 - 6)/31]$ to obtain the relevant number of degree days, and the number in the end month would be multiplied by $(5/30)$.

During the initial period, energy measurement periods of $365 \pm 31$ days (shown as $\pm n$ on Figure 2) will be accepted, but the OR calculation software will display a warning where the difference is greater than $\pm 7$ and less than $\pm 31$ days. Where the main heating energy is electricity, the maximum accepted discrepancy is $\pm 15$ days (shown as $\pm n$ on Figure 2) and the warning generated where the difference is between $\pm 7$ and $\pm 15$ days.

Similarly, the start or end dates of the energy measurement periods of fuels other than the main heating fuel must be aligned with the start and end dates of the assessment period within $\pm 31$ days (shown as $\pm m$ on Figure 2) or these will not be accepted. The OR calculation software will display a warning where the discrepancy in alignment of the measurement periods is between $\pm 7$ and $\pm 31$ days.

Where energy consumption measurement periods, or utility company estimate periods, are provided that are outside the allowed tolerances for measurement period length or alignment with that of the main heating fuel, then the DEC will show a default OR and other comparative details equivalent to those of a building with twice the energy consumption of one considered typical for its type. This will, by definition, equate to an OR of 200, and an automatic worst 'G' rating.

Figure 2 shows the principles of the allowable tolerances on energy measurement period length, the alignment of the assessment period with either the start or the end of the measurement period for the main heating fuel, and the alignment of other energy measurement periods with the assessment period.

The energy assessor should examine the start and end dates of all the measured energies, or utility estimates of energy use, that are available, before setting the assessment period to align with the start or end date of the main heating fuel measurement period. This may help to determine whether one option is preferable to the other, as one alignment option may bring the measurements within tolerance and the other may not.

### 5.3.4    On-site renewables and other low and zero carbon technologies

Low and zero carbon (LZC) technologies include on-site renewables (OSR), and may be defined as those technologies that supply energy to the building whilst reducing the metered energy supplied to the building from traditional sources. LZC technology therefore includes plant such as biofuel boilers and heat pumps.

If there is on-site generation of electricity from an LZC technology, this reduces the grid electricity demand (although CHP will increase the heating fuel demand). Similarly, solar thermal heating reduces heating fuel demand. Meter readings are not adjusted, but it is good practice to meter all LZC outputs so that the efficiency of the building itself can be assessed, and any carbon emissions thereby avoided may be credited on the DEC.

To be included in the calculation and acknowledged on the DEC, each OSR or LZC output energy measurement period must be within the range $365 \pm 15$ days where the output is electricity and $\pm 31$ days for other energy outputs, and the measurement period must be aligned to begin within $\pm 31$ days of the beginning, or end within $\pm 31$ days of the end, of the assessment period. The software that carries out the OR calculation will display a warning where the discrepancy in measurement period is between $\pm 7$ and $\pm 31$ days.

### 5.3.5    Carbon dioxide conversion factors

Energy use in buildings causes the emission of $CO_2$. For comparisons between buildings, standard $CO_2$ conversion factors are needed for each form of energy used in buildings. The $CO_2$ conversion factors used for calculation of ORs and production of DECs are one of a number of resources held and maintained by the national register on behalf of government on the Central Information Point (CIP). The current CIP data file is accessible on the national register website (http://www.ndepcregister.com) to all accredited energy assessors.

## 5.4    Typical carbon emissions (benchmarks)

Building performance can only usefully be compared with other buildings that perform similar functions. It is not helpful to compare, for example, an office with a hospital,

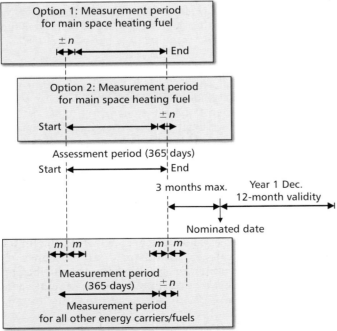

**Figure 2** Principles of aligning measurement periods and the assessment period

and so different performance benchmarks are required for each type of building function.

CIBSE has prepared operational benchmarks for 29 main categories of building, and have listed together the different types of building and use that would be included within each of the general category descriptions[15]. This is a limited number of categories. The number has been limited because the purpose of the Operational Rating and the DEC is to provide transparent information about building energy performance, and to motivate improvements. Providing a very large number of specialised benchmarks would not serve this purpose, as there would be numerous calls for special categories for buildings with particularly high energy use. This would limit the comparison of performance. An analogy is the use of the miles per gallon metric for cars. It is widely understood, and applies across the whole industry. Whilst there may be some interest in showing that one high-consumption vehicle is marginally less high-consuming than another, the system identifies both as being high-consumption.

The benchmarks are expressed in terms of energy density (kW·h/m$^2$ per year), and are separated into electrical and non-electrical components. Representative emissions densities (kgCO$_2$/m$^2$ per year) are also indicated, using representative CO$_2$ emission factors; the emission densities are intended for information only and are not used in the calculation procedure.

The benchmarks have been prepared to represent building use under standardised conditions, as follows:

— The weather year is standardised at 2021 degree-days per year, to a base temperature of 15.5 °C.

— A defined occupancy period is noted for each category individually.

— A standard proportion of the non-electrical energy density benchmark that is considered to be related to the heating demand is noted for each building category individually.

The purpose of these factors is explained below, as they may form part of the procedure of adjusting the benchmarks to represent better the characteristics of the building being assessed, where the basic benchmark should be adjusted to the location and use of that building.

### 5.4.1 Adjusting the benchmark for location (weather region)

The category benchmark is adjusted according to the 'history' of temperatures in the building location, for the one-year assessment period over which the OR is to be calculated. The adjustment is based on the number of monthly degree-days over the 12-month assessment period for the region in which the building is located. The monthly degree-day information is obtained from the Central Information Point, using the postcode of the building to determine the weather region appropriate for the building.

Where the assessment period does not include complete months at its beginning and end, then the procedure described in section 5.3.3 is applied.

For a building category with an 'electrical' energy density benchmark of $E$ (kW·h/m$^2$ per year) and a 'non-electrical' energy density benchmark of $N$ (kW·h/m$^2$ per year), the CO$_2$ density benchmark is found using standard CO$_2$ intensity factors obtained from the Central Information Point as follows (the non-electrical fuel is gas in this example):

$$C = (C_E E) + (C_G N) \tag{5.1}$$

where $C$ is the CO$_2$ density benchmark (kgCO$_2$/m$^2$ per year), $C_E$ is the CO$_2$ intensity factor for electricity (kgCO$_2$/kW·h), $E$ is the energy density benchmark for electrical energy (kW·h/m$^2$ per year), $C_G$ is the CO$_2$ intensity factor for gas (kgCO$_2$/kW·h) and $N$ is the energy density benchmark for non-electrical energy (kW·h/m$^2$ per year).

Only that part of the energy density benchmark that is related to heating demand is adjusted for the number of degree-days, i.e:

$$N_{dd} = [N(1 - P/100)] + [(N P/100)(L/S)] \tag{5.2}$$

where $N_{dd}$ is the thermal density benchmark adjusted for degree-days (kW·h/m$^2$ per year), $N$ is the non-electrical energy density benchmark (kW·h/m$^2$ per year), $P$ is the proportion of the non-electrical benchmark related to heating (%), $L$ is the number of degree-days in the assessment period for the specific location (degree-days) and $S$ is the standard degree-days for the category (degree-days).

The resulting 'degree-day corrected' CO$_2$ density benchmark becomes:

$$C_{dd} = (C_E E) + (C_G N_{dd}) \tag{5.3}$$

where $C_{dd}$ is the CO$_2$ density benchmark adjusted for degree-days (kgCO$_2$/m$^2$ per year), $C_E$ is the CO$_2$ intensity factor for electricity (kgCO$_2$/kW·h), $E$ is the energy density benchmark for electrical energy (kW·h/m$^2$ per year) and $C_G$ is the CO$_2$ intensity factor for gas (kgCO$_2$/kW·h).

The approved operational rating calculation methodology only allows for heating degree-days adjustments. Cooling degree-days are currently not included in the calculations.

### 5.4.2 Adjusting the benchmark for longer hours of occupancy

Where the energy assessor can demonstrate that the building is occupied for significantly longer periods than the standard hours quoted for the category, and where the benchmark data allow correction for extended hours of use to be made, then the degree-day corrected benchmark may also be adjusted for extended occupancy within the limits defined for that category. Adjustments for occupation beyond the maximum hours specified are not possible, as there is a limit to the marginal effect of extended operation on energy use.

Suitable forms of evidence to support the extended occupancy hours are detailed below and in section 4. These are required because of the risk that occupancy adjustments could be used to obtain better ORs than are justified by the standard methodology. Where available, the benchmark information includes separate correction factors for occupancy period for the electrical and the non-electrical energy density benchmarks. The benchmark information contains:

— benchmark (standard) hours per year: $H_B$ (h)

— maximum allowed hours per year: $H_{max}$ (h)

— percentage increase in electrical energy density benchmark at maximum allowed hours per year: $\Delta E$ (%)

— percentage increase in fossil-thermal energy density benchmark at maximum allowed hours per year: $\Delta P$ (%).

Where, for example, it can be shown that the actual use of a building is $H_a$ hours per year, then electrical and non-electrical energy densities corrected for both degree-days and occupancy may be calculated as:

$$E_{dd/occ} = E \{1 + [(H_a - H_B)/(H_{max} - H_B)] (\Delta E/100)\}$$

$$(5.4)$$

where $E_{dd/occ}$ is the electrical energy density adjusted for degree-days and occupancy (kW·h/m² per year).

$$N_{dd/occ} = N_{dd} \{1 + [(H_a - H_B)/(H_{max} - H_B)] (\Delta P/100)\}$$

$$(5.5)$$

where $N_{dd/occ}$ is the non-electrical energy density adjusted for degree-days and occupancy (kW·h/m² per year).

*Note*: $H_a$ must not exceed $H_{max}$.

The resulting $CO_2$ density benchmark corrected for degree-days and occupancy becomes:

$$C_{dd/occ} = (C_E E_{dd/occ}) + (C_G N_{dd/occ}) \qquad (5.6)$$

where $C_{dd/occ}$ is the $CO_2$ density benchmark adjusted for degree-days and occupancy (kgCO₂/m² per year).

To obtain annual occupancy hours the energy assessor uses the appropriate occupancy measurements indicated for each benchmark category. The two systems of defining annual occupancy hours are:

— the number of hours per year that the number of recorded occupants exceeds 25% of the usual maximum occupancy for that category of building

*or*:

— the number of hours per year that the premises are fully open to the public according to published opening hours (which may vary on a seasonal basis).

The energy assessor must obtain attendance records, survey results or published opening hours and calculate the annual occupancy hours. This information is to be collated into an 'annual occupancy hours record' and signed by the occupants' premises manager before the energy assessor uses the occupancy data in the Operational Rating calculation procedure. This record should be retained by the assessor for quality assurance purposes.

Where parts of the building falling within the same benchmark category have different occupancies the lowest occupancy must be used. Alternatively an assessment of occupancy in each part may be made and the occupancies combined using the percentages of overall floor areas, i.e. an area-weighted average occupancy.

Where occupancy adjustment is required in a multiple-use building falling into more than one benchmark category, the annual occupancy hours must be calculated as above for each category for which an occupancy adjustment is relevant.

### 5.4.3 Mixed-use assessment: composite benchmarks

A composite benchmark may be needed where the activities taking place in the occupied space being assessed fall into more than one of the building categories for which there are separate benchmarks, e.g. an office with an integral leisure centre, a library within a civic centre for performing arts.

Where each activity area exceeds 1000 m², and is separately metered, then it may be appropriate to assess each area separately and produce a separate DEC for each area. In some situations, where the energy for each activity is separately metered, a separate DEC may be produced (on a voluntary basis) for each activity, even if the area for any activity is less than 1000 m². Alternatively, and where the separate activities are not separately metered, an overall DEC can be produced using a composite benchmark based on an area-weighted average of the individual benchmarks.

This is calculated as follows, where:

— $E_{dd/occ}(a, b, ... n)$ is the electrical energy density corrected for degree-days and occupancy associated with area (a, b, ... n)

— $N_{dd/occ}(a, b, ... n)$ is the non-electrical energy density corrected for degree-days and occupancy associated with area (a, b, ... n)

— $C_{dd/occ}(a, b, ... n)$ is the $CO_2$ density benchmark adjusted for degree-days and occupancy associated with area (a, b, ... n)

— $A(a, b, ... n)$ is the floor area of area (a, b, ... n).

Then the composite $CO_2$ benchmark for electrical energy is given by:

$$\text{CBM}_E = \{[E_{dd/occ}(a) A(a)] + [E_{dd/occ}(b) A(b)] + .....$$
$$+ [E_{dd/occ}(n) A(n)]\} / [A(a) + A(b) + ... + A(n)]$$

$$(5.7)$$

where $\text{CBM}_E$ is the $CO_2$ benchmark for electricity (kW·h/m² per year) and $E_{dd/occ}(a)$ is the electrical energy density adjusted for degree-days and occupancy (kW·h/m² per year) associated with area $A(a)$ etc.

Then the composite $CO_2$ benchmark for non-electrical energy is given by:

$$\text{CBM}_N = \{[N_{dd/occ}(a)\,A(a)] + [N_{dd/occ}(b)\,A(b)] + .....$$
$$+ [N_{dd/occ}(n)\,A(n)]\} / [A(a) + A(b) + ... + A(n)]$$

$$(5.8)$$

where $\text{CBM}_N$ is the $CO_2$ benchmark for electricity (kW·h/m² per year) and $N_{dd/occ}(a)$ is the non-electrical energy density adjusted for degree-days and occupancy (kW·h/m² per year) associated with area $A(a)$ etc.

The overall composite $CO_2$ benchmark is given by:

$$\text{CBM} = \{[C_{dd/occ}(a)\,A(a)] + [C_{dd/occ}(b)\,A(b)] + .....$$
$$+ [C_{dd/occ}(n)\,A(n)]\} / [A(a) + A(b) + ... + A(n)]$$

$$(5.9)$$

where CBM is the composite $CO_2$ benchmark (kgCO₂/m² per year) and $C_{dd/occ}(a)$ is the $CO_2$ density benchmark adjusted for degree-days and occupancy associated with area (a) (kgCO₂/m² per year) etc.

*Example*

In a school with a swimming pool, that part of the school excluding the swimming pool would be assigned the benchmark for 'schools and seasonal public buildings' (category 17). The swimming pool would be assigned the benchmark for a 'swimming pool centre' (category 12). The composite benchmark is the area-weighted average of the benchmarks for the two categories, i.e:

$$\text{CBM} = \frac{(\text{BM}_{school} \times A_{school}) + (\text{BM}_{pool} \times A_{pool})}{A_{school} + A_{pool}}$$

where CBM is the composite benchmark for the school including swimming pool (kgCO₂/m² per year), $\text{BM}_{school}$ is the benchmark for a school (excluding swimming pool) (kgCO₂/m² per year), $A_{school}$ is the floor area of the school (excluding swimming pool) (m²), $\text{BM}_{pool}$ is the benchmark for a school swimming pool centre (kgCO₂/m² per year) and $A_{pool}$ is the floor area of the swimming pool centre (m²).

## 5.5 Electrical and non-electrical energy ratios

The more comprehensive 'technical table' provides additional detail of the assessment. This includes a comparison between the relevant actual consumptions of electrical and non-electrical energy over the assessment period and electrical and non-electrical energy densities corrected for degree-days and occupancy ($E_{dd/occ}$ and $N_{dd/occ}$ respectively). These are expressed as the electrical ($R_E$) and non-electrical ($R_N$) ratios:

$$R_E = \frac{\text{Relevant electrical consumption}}{E_{dd/occ}} \qquad (5.10)$$

and

$$R_N = \frac{\text{Relevant non-electrical consumption}}{N_{dd/occ}} \qquad (5.11)$$

These factors are useful to energy assessors as they may indicate where resources need to be focussed in order to identify potential reasons for poor performance for inclusion in the advisory report. These ratios indicate where higher than expected energy demand is occurring, and this observation should alert the energy assessor to inefficiency in the operation of the building.

## 5.6 The Operational Rating

The Operational Rating (OR) is calculated as the relevant total carbon dioxide emissions of the building over the assessment period divided by the $CO_2$ density benchmark corrected for degree-days and occupancy ($C_{dd/occ}$). To avoid fractions, the result of the comparison is expressed as a percentage, rounded to the nearest whole number:

$$\text{OR} = (\text{building } CO_2 \text{ emissions} / \text{building area})$$
$$\times (100 / \text{typical } CO_2 \text{ emissions per unit area})$$

$$(5.12)$$

In the case of a composite benchmark assessment, the relevant total carbon dioxide emissions of the building over the assessment period are divided by the overall composite $CO_2$ benchmark (CBM), calculated as described in section 5.4.3.

Where any estimates of energy consumption have been used in the calculation of the OR, the DEC will show that the result has been estimated, see section 4.1.4.2.

## 5.7 The DEC rating

The DEC is required to display the performance of the building as a label. The Government has decided that this label will use an A to G scale, which is now familiar to the public through its use for household appliances.

The values of the Operational Rating for each grade are defined as shown in Table 2. The DEC also shows the month and year of the nominated date (see section 5.3.3) of the assessment.

**Table 2** Definition of Operational Rating bands

| OR band | Label |
|---|---|
| 0 to 25 | A |
| 26 to 50 | B |
| 51 to 75 | C |
| 76 to 100 | D |
| 101 to 125 | E |
| 126 to 150 | F |
| More than 150 | G |

## 5.8 Benefits of renewable energy sources and LZC technologies

A number of technologies are available that generate heat or electricity from ambient sources, with either very low or zero carbon emissions. They are often referred to as low or zero carbon technologies (LZCs), or as on-site renewables (OSRs). The DEC includes a histogram (Figure 1, item 'C'), which shows the total contributions to carbon dioxide emissions. It identifies three contributions to the total emissions: those from heating, those from electrical energy consumption and those from on-site renewable energy sources. This section describes what is included within the on-site renewables definition, and section 5.9 describes the heating and electrical elements of the carbon dioxide emissions. To be separately identified, any LZC must be separately metered.

Energy delivered to the building by low or zero carbon technologies reduces the metered consumption of energy from conventional sources. On-site renewables include solar thermal, geothermal, photovoltaic and wind energy systems. Low or zero carbon technologies include biofuel boilers, combined heat and power systems (CHP) and heat pumps. Any of these technologies may take the place of, or supplement, plant that would otherwise have been included as part of a conventional servicing strategy.

The light grey portion of the DEC histogram (Figure 1, item 'C') that is below the zero line indicates, for the current assessment year and for the two preceding assessment periods, where data are available:

— actual carbon dioxide emissions savings for on-site renewables

— the carbon dioxide emissions that would have been emitted by a conventional system such as a gas fired boiler.

Where both on-site renewables and low carbon technologies are installed, the grey portion of the bar, below the line, represents the aggregate of both of these.

Where the energy from on-site renewables or from a low carbon technology is specifically measured, within the limits to the length of measurement period and in alignment with the assessment period specified at 5.3.5 above, then the contribution to reducing the building's $CO_2$ emissions is calculated as follows.

Where the output from the on-site renewables or low carbon technology is electricity, then the $CO_2$ contribution is calculated as the electricity supplied in the 365-day period, multiplied by the $CO_2$ conversion factor for electricity obtained from the Central Information Point. This calculation is also be appropriate to calculate the $CO_2$ contribution from a metered supply of electricity from an on-site CHP installation, namely:

$$CO_2 \text{ contribution} = 365\text{-day electricity supplied}$$
$$\times CO_2 \text{ conversion factor (electricity)}$$
$$(5.13)$$

Where the energy output from the on-site renewables or low carbon technology is thermal (heating) energy, the $CO_2$ contribution is calculated as the $CO_2$ emissions that would have been produced by a boiler operating at a seasonal efficiency of 0.8 (the current minimum value for replacement boilers as defined in the CLG's *Non-domestic Heating, Ventilation and Air Conditioning Compliance Guide*[20]★) to provide the same thermal energy. The $CO_2$ contribution is:

---

★ This publication is under revision at the time of writing (May 2009).

$$CO_2 \text{ contribution} = (365\text{-day heating energy}/0.8)$$
$$\times CO_2 \text{ conversion factor (main heating fuel)}$$
$$(5.14)$$

However, where the building is heated using a heat pump, the $CO_2$ conversion factor for natural gas should always be used, even though gas may not be the main heating fuel.

Where the LZC output is cooling energy, the $CO_2$ contribution is calculated as the $CO_2$ emissions that would have been produced by a conventional electrically driven cooling plant operating at a seasonal efficiency of 2.25 (the current minimum energy efficiency ratio for an air cooled vapour compression chiller as defined in the CLG's *Non-domestic Heating, Ventilation and Air Conditioning Compliance Guide*[20] to provide the same cooling energy. In this case the $CO_2$ contribution is:

$$CO_2 \text{ contribution} = (365\text{-day cooling energy}/2.25)$$
$$\times CO_2 \text{ conversion factor (electricity)}$$
$$(5.15)$$

## 5.9    Total $CO_2$ emissions

The $CO_2$ emissions histogram on the DEC (see Figure 1, item 'C') indicates for the current assessment year and for the two preceding assessment periods (where data are available):

— contribution to the total, the annual $CO_2$ emissions due to the consumption of grid supplied electricity (above the line)

— contribution to the total, the annual $CO_2$ emissions due to the consumption of all other forms of delivered energy (above the line); for convenience this is termed 'heating' although it includes delivered fuels that are used for any purpose, together with any delivered heating and cooling energy from, for example, district heating or cooling systems

— any annual $CO_2$ contributions from on-site renewable energy sources and other LZC technologies (below the line).

The histogram thus indicates the total carbon emissions that would arise without any LZC (the whole of the bar) and the actual emissions arising (the part of the bar above the line.)

## 5.10    The technical table

The DEC itself displays a small table of 'technical information' in the lower left hand corner of the certificate (see Figure 1, item 'G'). This details:

— whether the energy consumption data used to compile the DEC was from actual meter readings or estimates provided by the utility supplier

— the main heating fuel used

— the predominant servicing strategy, e.g. naturally ventilated or air conditioned

— the total useful floor area

— where available, the asset rating.

In addition, it gives annual energy use for both heating and electrical energy, and the typical values for a building of that type, together with any energy from renewable sources, either on-site renewables or LZC technologies.

The calculation procedure that generates the OR and provides the results to be displayed on the DEC also provides a more comprehensive technical table that shows additional information about the assessment. This is not part of the DEC and is not subject to any regulatory requirements for either display or lodgement. It provides a fuller picture of the building's energy performance and will be of assistance in developing an action plan to improve the energy performance of the building. The energy assessor should make the technical table available to the occupier alongside the DEC and advisory report, and should use the table to help make recommendations in the advisory report.

The technical table is particularly important in buildings where there is a significant contribution from on-site renewables or LZC technologies. These buildings may exhibit low total carbon emissions, whilst using energy inefficiently, because a significant proportion of the energy used is from renewable or low carbon sources. Because the Operational Rating is expressed in terms of a ratio of overall carbon dioxide emissions relative to the benchmark, it is quite possible for a very poorly managed building to appear to be well run if it obtains a significant proportion of its energy from on-site renewables or from LZC sources. The information in the full technical table will enable the energy assessor to identify this and to advise the building occupier about appropriate remedial measures.

The technical table includes the following:

— total useful floor area (TUFA) (m$^2$)

— total unconditioned area (m$^2$)

— area (m$^2$) and category name of each separable energy use

— floor area used in the (following) performance calculations (TUFA minus the separable areas) (m$^2$)

— current Asset Rating and Asset Rating grade of the building (if available)

— totals of thermal and of electrical energy used in the assessment period (kW·h), and the resulting $CO_2$ emissions (tonnes $CO_2$)

— the thermal and electrical energy use attributed to each separable area (kW·h), and the resulting $CO_2$ emissions (tonnes $CO_2$)

— calculated thermal, electrical and $CO_2$ performance indicators (kW·h/m$^2$) and (kg$CO_2$/m$^2$) respectively

— reference thermal, electrical and $CO_2$ performance benchmarks (kW·h/m$^2$) and (kg$CO_2$/m$^2$)

— benchmark ratios for thermal and electrical energy performance

— benchmark ratio for $CO_2$ performance (the calculated Operational Rating), and OR grade

— the percentage of thermal and electrical energy provided from LZC systems

— the percentage of $CO_2$ emissions avoided by the use of LZC systems.

In these descriptions, it should be noted that:

— 'thermal' includes imported combustion fuels (such as fossil and bio-fuels) used for all purposes, and includes any heating and cooling energy from community systems

— 'electrical' includes electricity used for all purposes including heating and cooling

— the area used in the performance comparisons includes accessible unconditioned spaces, but excludes separable energy use areas.

The technical table is a very useful output from the certification process, and energy assessors should be conversant with the information it conveys when preparing the advisory report.

The technical table contains information that is not included in the formal DEC because the CLG wished to keep the DEC as simple as possible to aid wider public understanding and acceptance. However, it contains valuable indicators that show how the building is using energy.

For example, a building using a low carbon heating source such as a biomass boiler may achieve a good OR and DEC grade as this is related to the carbon intensity, but it may be using more low carbon fuel than necessary. Only the technical table would fully reveals this.

In the hands of the energy assessor, the technical table should be a vital tool in producing meaningful recommendations tailored to the specific building, and therefore a key aid to delivering appropriate advice to assist the client occupier to save energy, money and carbon.

## 5.11     The Central Information Point

A number of the factors needed to calculate the Operational Rating must be standardised and kept up to date so that all energy assessors make use of the same approved data. The Government provides these through a Central Information Point (CIP) data file available to accredited energy assessors from the national register website (http://www.ndepcregister.com). Accredited energy assessors should download the necessary data file to enable the approved software to calculate the OR, and provide the information to produce the DEC and the advisory report. The contents of the CIP are updated every month and include:

— the approved benchmark data, providing the electricity and non-electrical energy densities (benchmarks), the reference degree-days and approved conversion and adjustment factors for the building categories, and a tabulation showing how indicative building types are 'allocated' to the building benchmark categories

— the approved monthly degree-day information for the weather regions of the UK; data for the three year period up to the most recent month are accessible directly, with records maintained for a ten year period for quality assurance purposes

— a table allowing the appropriate weather region to be obtained from the postcode of the building's location

— a list of currently approved accreditation schemes.

# 6    Advisory reports

Regulation 16(2) (as amended) states that, 'except where Regulation 18(3) applies, on and after 1 October 2008 every occupier of any building to which this regulation applies must—

(a) have in its possession or control at all times a valid advisory report; and

(b) display at all times a valid display energy certificate in a prominent place clearly visible to the public.'

Regulation 19 states that 'an advisory report is a report issued by an energy assessor after his assessment of the building, which contains recommendations for improvement of the energy performance of the building.'

This section considers the requirements for an advisory report, what information is needed to produce the report, how to use the report generator software tool, and how existing energy survey reports can be used.

An advisory report must be produced at the same time as the first display energy certificate (DEC). A lodged DEC is not legally valid without the advisory report also being lodged. Although the report does not have to be publicly displayed, responsible persons must have it in their possession or control at all times. As with the DEC, the advisory report is produced in a standard format and using an approved software tool. A copy could be requested under the Freedon of Information Act[13].

The software tool is based on a master list of standard recommendations for improvement which are listed in the software specifications and held on the software. These recommendations have been assembled from data made available by the Carbon Trust and from key publications such as CIBSE Guides, other CIBSE publications and Good Practice/Energy Efficiency Guides.

There is an expectation that to prepare the advisory report, the energy assessor must undertake a 'walk-round' energy survey of the building. CLG has recently indicated that it is expected that the assessor will visit the building. Where an existing advisory report or recommendation report is available, the energy assessor should establish with the building occupier the actions planned or completed in response to the previous report.

Based on information about the building provided by the energy assessor, the tool automatically filters the generic recommendations into a shortlist that is applicable to the building being assessed. Any recommendations that are inappropriate for the building, and are not automatically filtered out, must be manually removed by the energy assessor to ensure the accuracy and appropriateness of the report. The energy assessor should then prioritise the

recommendations according to their likely potential to reduce the carbon emissions of the building. Carbon impacts are grouped in bands showing which measures are likely to have 'high', 'medium' or 'low' impact on overall building emissions.

Simple payback periods for each measure have been estimated using available data for all building types. Consequently, reported payback may not accurately reflect actual payback for the building being assessed.

The building owner or occupier may possess an earlier energy audit or report relating to the building. Where this is available, the energy assessor may decide to use information from that report in preparing the advisory report. However, the energy assessor must establish the accuracy of such reports, as he/she takes full responsibility for any elements used to compile the advisory report.

Additionally, the energy assessor may enter additional recommendations that are particularly applicable to the building into a 'free text' box. Any such recommendations should originate from a sound source such as a recent building EPC recommendations report, Low Carbon Consultant or Carbon Trust energy survey report, air conditioning inspection report, boiler energy efficiency inspection report, or any other report prepared by a suitably qualified professional. The energy assessor is responsible for deciding whether to include any such recommendations.

The energy assessor's accreditation scheme quality assurance system may require the energy assessor to justify any deletions, re-rankings, carbon impact assessments or additional measures recommended. Therefore the energy assessor should keep accurate records of the decisions taken.

Surveys and inspections are required by other parts of the Regulations:

(a)    Part 2 covers EPCs for construction sale or rent, which are accompanied by a recommendations report listing cost-effective measures. Since these certificates are principally about the performance of the asset, the list of measures accompanying the EPC may give greater emphasis to building improvements as opposed to behavioural advice. For further information, see *Energy and carbon emissions regulations*[11].

(b)    Part 4 covers air conditioning system inspections. The reports generated under these inspection regulations will give advice on cost effective improvements to the system. It should be noted however that the maximum periods between advisory reports and between air conditioning inspection reports is different (i.e 7 years as opposed to 5 years). For further information, see CIBSE TM44: *Inspection of air conditioning systems*[21].

These surveys and inspections must be carried out by a 'qualified and/or accredited expert' and the energy assessor must visit the site. It may be that a single energy assessor has the appropriate qualifications and is accredited to carry out more than one of the required tasks. Consequently, building owners may find it cost-effective to schedule work so that one site visit can generate the information to satisfy more than one of the requirements of the Regulations.

## 6.1 Collecting information for advisory reports

This section describes the information that the energy assessor should collect for entry into an approved advisory report software tool. Supplying this information enables the tool to filter the database of standard measures down to a shortlist of recommendations which are likely to be applicable to the building. This filtered list may be further refined by the energy assessor.

There is a requirement that a site survey is undertaken to gather information relating to the building and its energy systems. The energy assessor may also use prior knowledge of the building for their report. Such prior knowledge may have been gathered by undertaking building energy management duties, producing or reviewing existing energy reports or asset management plans. Where these are used, the energy assessor is responsible for ensuring that the advisory report contains appropriate recommendations.

## 6.2 Building elements to be considered

When surveying the building, or reviewing existing information for the purposes of preparing an advisory report, the following elements of the building should be considered.

### Building fabric

Building fabric relates to the condition and thermal performance of building elements such as roofs, walls, windows, doors and floors. The energy assessor must have a basic knowledge, gather evidence or assess evidence provided relating to the type of construction and its condition to answer this section.

### Control of heating, ventilation and air conditioning

Control of heating, ventilation and air conditioning (HVAC) relates to the control and operation of building services used to heat, cool and ventilate the building. This includes the physical control systems in place and the way in which they are managed. The energy assessor should seek evidence of the current settings and management practices in place to ensure they are regularly checked and adjusted to suit building occupancy.

### Heating systems

Basic information about the heating systems must be established to proceed with this section. Evidence should be sought in respect of planned inspections and servicing regimes in place. Every effort should be made to locate and use existing boiler inspection reports, and visually inspect the heating plant for condition, leaking, insulation and corrosion.

### Ventilation

The energy assessor must establish whether the building has natural, mixed mode ventilation or full air conditioning. The energy assessor then need only answer questions pertinent to the type of system in place. Having established the strategy, the energy assessor should check whether elements of the system operate appropriately (e.g. operable windows that are locked for security reasons or air handling equipment that shows no evidence of adequate cleaning and maintenance). The energy assessor may seek evidence of whether the ventilation strategy is adequate to maintain the required indoor environment, in particular evidence of summertime overheating.

### Air conditioning systems

Information about the age, condition and maintenance regimes of air conditioning plant should be sought. It would be advantageous for the energy assessor to obtain existing air conditioning inspection reports for information.

### Lighting

Lighting covers both electric and natural internal lighting and external lighting systems. The energy assessor should establish what maintenance regimes are in place and if the lighting strategy has been recently reviewed to determine whether it matches current needs.

### Domestic hot water

Basic information about the domestic hot water system (DHWS) should be obtained. Evidence should be sought in respect of water saving devices fitted and the condition of the DHWS plant.

### Occupier's energy consuming equipment

Evidence should be sought as to how effectively energy consuming equipment such as personal computers, printers, faxes, portable heaters/air conditioners and vending machines is used. Are users encouraged to switch their own equipment off? Are power-save settings used? Are sufficient automated controls in place, etc?

### Lifts and escalators

Energy assessors should establish what metering is in place for lifts and escalators, and if alternative methods of travelling between floors are available to the building users.

### Alternative energy

The energy assessor should make a subjective judgment on what renewable energy sources and low and zero carbon technologies may be suitable for the building. The decision to include particular measures should be based on knowledge of the building form, fabric and energy demands.

## Swimming pools

The energy assessor should establish what energy metering is in place for the pool. Are pool covers installed and used effectively? Is the pool hall effectively insulated and isolated from the atmosphere by, for example, air-locked doors?

## Catering

For stand-alone restaurants and for buildings that have a commercial catering facility (not including staff kitchens), it is necessary to review the current energy management practices, energy metering and the condition, suitability and use of equipment.

## Specialist equipment

The type of equipment will vary from site to site but will typically include fume cupboards and specialist process equipment such as that found in laboratories, etc. Manufacturers' guidance, condition survey reports, maintenance reports etc. should be sought and a visual inspection undertaken to ascertain whether the equipment is energy efficient and is being used adequately.

## Steam systems

Evidence should be sought in respect of planned inspections and servicing regimes for the steam plant. Every effort should be made to locate and use existing steam plant inspection reports and to visually inspect the equipment for condition, leaking, insulation and corrosion.

## Operation and management

Seek evidence of current energy management practices. Areas of interest include the operation of building services, management of spaces and staff interaction, responsibility and guidance. Evidence might be sought by speaking to site representatives with appropriate knowledge to comment. It is necessary to establish what meters are in place and whether this provision is sufficient. Newer buildings completed under the 2001 amendments to Part L of the Building Regulations[18] should have a building log book and should have full commissioning records. Further guidance on these is available in CIBSE TM31[22] and the CIBSE Commissioning Code M[23].

## 6.3     How to use the advisory report software

This section describes the questions asked by the advisory report software tool, which should be answered using the information gathered as described in section 6.1. Some information, such as whether the building is 'listed' or has a previous advisory report, should be addressed to the client in the first instance, since they should have this information. Where information is not available the energy assessor may answer 'not available'. However, overuse of this approach may create some concerns in relation to the accreditation scheme's quality checking of the advisory reports. The questions are as follows.

### 6.3.1     Building background information

*Report/survey type*

The energy assessor is asked to confirm that a site survey has been carried out.

*Previous advisory reports*

It is necessary to confirm the existence of any previous advisory report via the Landmark® database[14]. If a previous advisory report has been produced it is suggested that the energy assessor checks the occupier's actions, planned or completed, in response to that earlier report.

*Listed buildings*

The energy assessor must establish if the building is listed. If it is, the advisory report will include a link to the English Heritage website (http://www.english-heritage.org.uk) providing specific measures for historic buildings. The CIBSE's *Guide to building services for historic buildings*[24] is also relevant.

*Special energy use(s)*

If any part of the site has been discounted in the calculation of the DEC as separable energy uses, such as regional data centres, pools, commercial catering etc., the energy assessor should indicate it here.

*Building features*

This is the first stage of the recommendations filtering process. The energy assessor must select specific building features and energy uses from a list. Selecting these items will ensure the relevant sections of the questionnaire are presented to the energy assessor.

### 6.3.2     Specific building features questions

A new tab appears across the top of the software screen for each of the specific building features that is selected. Each tab includes a series of questions, either as drop-down lists that identify specific improvement measures, or to eliminate recommendations that are not relevant to the building being assessed. Some questions offer multiple-choice answers, usually following the format 'yes', 'no' or 'don't know'. If 'don't know' is selected, this will identify default recommendations, for example prompting the energy assessor to seek further information.

The energy assessor can answer each section in any order. Unanswered questions are flagged and must be answered before proceeding to the next stage, which produces the shortlist of selected recommendations. Work can be saved and returned to in a subsequent session if required. Once all relevant questions have been answered the energy assessor moves on to the next stages to delete inappropriate recommendations, identify carbon impact, and make additional recommendations.

Recommendations are displayed in three categories:

— *Table 1*: no cost and low cost (0 to 3 years payback)

— *Table 2*: medium payback (3 to 7 years payback)

—    *Table 3*: strategic investment (7+ years payback).

Once the shortlist of recommendations appears, divided into the three payback categories, the energy assessor can remove unsuitable recommendations from the list based on his/her knowledge of the site.

In order to give the recipient of the advisory report some indication as to which of the recommendations are more or less likely to affect the carbon emissions of their building, the energy assessor is asked to give an estimate of the carbon impact of each of the measures selected for inclusion in the advisory report.

A maximum of 30 standard recommendations may be selected for inclusion in the advisory report. In most instances the top 30 recommendations are likely to be low- and no-cost. For those occupiers that would like to consider investing to reduce their carbon emissions, it is important to ensure the advisory report is not dominated by these low- and no-cost recommendations. The software will therefore allow up to 30 recommendations to be selected, spread across the three simple payback categories as follows:

—    *Table 1*: no-cost and low-cost (0 to 3 years payback); the top 15 recommendations

—    *Table 2*: medium payback (3 to 7 years payback); the top 10 recommendations

—    *Table 3*: strategic investment (+7 years payback): the top 5 recommendations.

Changes may be saved at any stage of the advisory report generation process for safe keeping or to return to in a subsequent session. As with any other software, it is good practice to save changes regularly throughout the generation of the advisory report to reduce the risk of losing any work.

## 6.4      Report production

Once the energy assessor has finalised the list of recommendations, the software then populates the advisory report template and formats it for lodgement.

## 7      Registration and accreditation of energy assessors

Display Energy Certificates and advisory reports can only be issued by energy assessors who are accredited by a scheme approved by the Secretary of State. There are several categories of accredited energy assessor:

(1)    domestic energy assessors (DEAs)

(2)    on-construction domestic energy assessors

(3)    energy assessors for DECs and advisory reports

(4)    energy assessors for non-dwellings to produce EPCs and recommendation reports

(5)    air conditioning inspectors.

The accreditation schemes must be approved, and the energy assessors must be accredited, separately for each of these categories, and may only operate in the categories for which they have been approved or accredited.

Energy assessors must demonstrate that they either have an approved qualification as an energy assessor, or that they fulfil the 'approved prior experiential learning' (APEL) criteria for direct entry on the basis of demonstrable knowledge and experience. They must complete an 'APEL' form to describe their experience and demonstrate how it meets the requirements for knowledge and understanding set out in the CLG's *Minimum Requirements for Energy Assessors for Public Buildings (Display Energy Certificates)*[25]. They must also demonstrate three example DECs to show their competence with the software tool ORCalc or an approved alternative tool.

Energy assessors must maintain their registration with an accreditation scheme, which is responsible for the maintenance of standards and verifying the competency of accredited energy assessors. The accreditation scheme is also responsible for lodging certificates and reports produced by the energy assessors accredited under their scheme. The accreditation schemes lodge the data files containing the certificate and/or advisory report with the national register to ensure that every building assessment is referenced and available for download to the public.

Certificates and reports lodged onto the national register are also accessible to local authority trading standards departments to enable enforcement of the Regulations and to central government to enable statistical analysis and benchmarking. Energy assessors should submit the data file containing the DEC and advisory report to their accreditation scheme for quality assurance and lodgement on the central register before making the certificate and report available to the building occupier. On receipt of a certificate and report from an energy assessor, occupiers may wish to log-on to the national register website (http://www.ndepcregister.com) to verify that the certificate and report received have been duly lodged.

## 8      Landlord's Energy Statement and Tenant's Energy Review

The principle of measuring energy consumption in order to inform building users about the potential for them to improve their use of energy is straightforward. However, in buildings which have multiple tenants the practical details are far from straightforward. Yet such information is required to enable public authority tenants in buildings having multiple occupants to comply with obligations under the Regulations. In addition, the proposals for the Carbon Reduction Commitment[26] create additional reporting requirements that go well beyond the public sector, and affect many private landlords and their tenants.

To address the need for information about energy use in multi-tenanted buildings, the British Property Federation, the Usable Buildings Trust, CIBSE and the British Council for Offices, with funding from the Carbon Trust, have joined forces to develop the Landlord's Energy Statement (LES), and the Tenant's Energy Review (TER). These tools enable landlords and tenants to identify the energy being used in the building, and who is using it.

## 8.1 What is the Landlord's Energy Statement?

The Landlord's Energy Statement (LES) is a mechanism to enable landlords to identify, understand and report on the energy used by the common services (such as heating, ventilation and lifts) in their buildings, and the resulting $CO_2$ emissions. Specifically, it

—   identifies what energy is being used and where

—   takes account of the forms of energy being used

—   converts the energy used into carbon dioxide emissions

—   allows the energy efficiency of landlord's services to be compared with sectoral benchmarks to produce ratings

—   allows year-on-year performance improvements to be tracked

—   provides a robust energy reporting system

—   takes separate account of both on-site and off-site renewable energy supplies

—   gives tenants who need to prepare display energy certificates the necessary information on the energy and carbon content of their landlord's services.

The Landlord's Energy Statement has been developed primarily as an output specification for the property industry, which is designed to allow reports to be generated from data already held in landlords' and managing agents' computer systems. A Microsoft® Excel workbook is available, which helps the landlord to collect the required data, calculates the energy efficiency performance and carbon dioxide emissions and compares the performance of the landlord's services with industry benchmarks.

For buildings with multiple tenants, the tool offers a choice of completing an LES for the whole building landlord services for all occupants, or a tenant LES prepared specifically for each tenant, showing how much energy the landlord used on the individual tenant's behalf during the year.

## 8.2 What is the Tenant's Energy Review?

It assists tenants in making energy savings through the production of an automated energy savings report, helping to identify potential savings opportunities.

The Tenant's Energy Review (TER) is a tool which allows tenants to assemble information on the energy and emissions attributable to the energy they purchase directly from utility companies. It enables them to quantify direct energy use and the features likely to influence energy demand. It also identifies issues such as high occupation densities or long hours of use, as well as areas of intensive energy use within the tenancy such as dealing rooms and data centres, which all increase the energy consumption of the building. The TER uses the data collected to generate an energy savings report which identifies potential reductions in the use of electricity from direct supplies

within the tenant's demise and any other opportunities to improve energy performance.

The TER will help a tenant to better understand their energy use and to identify improvements. It will also serve as the basis for briefing any experts they may wish to engage to assist them in improving their energy performance. Together with the LES, the TER enables tenants to collate the information needed for a Display Energy Certificate (DEC), whether this is a voluntary exercise or a statutory duty for the tenant.

A further benefit of the LES-TER approach is that it enables greater mutual understanding and benchmarking of energy performance in multi-tenanted buildings, and supports landlords with larger portfolios in identifying the relative performance of buildings across their portfolio.

## 8.3 LES-TER and the Carbon Reduction Commitment

Introduction of the Carbon Reduction Commitment[26] in 2010 will require large organisations to report their overall annual $CO_2$ emissions, whilst also providing an incentive to improve year-on-year performance. Due to the nomination of the holder of the energy supply contract as the responsible CRC participant, many landlords will become responsible for their tenants' emissions. The LES-TER approach will provide a useful mechanism to help improve year-on-year performance for landlords and tenants affected by the Scheme.

Use of the LES and TER tools is free; having registered their details, users are then able to download the workbooks and guidance from the website (http://www.les-ter.org).

# 9   Frequently asked questions

The following questions have been asked on a number of occasions. The answers given are correct in the view of the contributors, but readers should be aware that only the courts can give a definitive answer to any question relating to the interpretation of Regulations.

Readers should be aware that neither CIBSE, nor any other accreditation scheme, nor CLG officials can offer definitive rulings on the meaning or interpretation of the Regulations, and should use the material given below and throughout this document accordingly.

## 9.1 DECs for new buildings

Q1   My client has just occupied an office over 1000 m² for the first time. I do not have 12 months' worth of meter readings for this client, so how can I produce a DEC? I have heard suggestions that if you don't have sufficient data then a 'G' rating of 200 is given. Or should I use the data from the previous occupant?

A1   You should not use readings from the previous occupier, nor are you forced to take a 'G' rating.

Under Regulation 18 you must display a DEC, but during the first fifteen months of occupancy this should show the Asset Rating of the building, and the Operational Rating is left blank, as are the bar graphs. This is known as an 'EPC-only DEC'.

Q2    What is the correct procedure for entering data and producing a DEC for a new building? What do I need to do in the software to generate the blank Operational Rating?

A2    In new buildings, there will not be any qualifying meter readings or fuel measurements available until it has been occupied for more than a year, so the only DEC that can be produced is an 'EPC-only DEC', see section 2.13.

In ORCalc, the user fills in the information as usual, up to the 'Past DEC Data' page. Here there is a box labelled 'Valid EPC Information'. Enter the EPC rating for the new building and tick 'Generate Asset Rating Only DEC'. When the user presses the 'Next' button, the software will jump straight to the 'Air conditioning Survey' page (bypassing the pages in between).

Q3    I have produced an EPC-only DEC. Do I need to produce an advisory report? Will the software allow me to do this when there is no Operational Rating?

A3    You should produce an advisory report. Even without an Operational Rating the software tool is still able to produce an advisory report for such a condition.

## 9.2    Software

Q1    I have completed a DEC but the lodgement has failed. Why is this?

A1    The lodgement process includes various checks on the content of the DEC. Simple errors such as zeros instead of 'O's (or vice versa), spaces where there should not be spaces, or missing data can cause lodgement to fail. Care should be taken when compiling a DEC or an advisory report to ensure that the fields are correctly entered.

Another cause of failed lodgements relates to incorrect insurance details. The software automatically checks that you have current insurance and will not allow you to lodge without it. Some assessors have found it necessary to accelerate the insurance renewal process in order to address this. Assessors working for public bodies may wish to clarify the situation with respect to their insurance arrangements.

Q2    I cannot find the correct address details, post code or UPRN.

A2    Submit details to Landmark® through your accreditation scheme to request the Property Address and UPRN on the Central Register.

Q3    Is there a helpdesk for ORCalc questions?

A3    Neither CLG nor their contractors for ORCalc offer a public help facility. Questions have to be submitted to the assessor's accreditation scheme. If the scheme is unable to provide a satisfactory answer, then they will refer the question to CLG or their contractor for attention.

Assessors should not contact CLG or their contractor directly, as they will be referred back to the accreditation scheme.

Q4    Do I have to use ORCalc to produce a DEC?

A4    There are a number of software packages approved by CLG for the purpose of producing DECs. A full list of those which are currently approved can be found at http://www.ukreg-accreditation.org/Index.html

## 9.3    Benchmarks

Q1    My building is open 24/7, but there is no allowance for this. What should I do?

A1    Use the maximum permitted occupancy for the category of building. Occupancy poses a particular problem. Allowing people to claim extended occupancy is probably the single largest opportunity for people to cheat with DECs, and so the use of occupancy adjustments is very tightly controlled. In addition, there is a limited evidence base for the real impact of extended occupancy hours on energy consumption – in some cases there is a large energy load if the building opens at all (a swimming pool, for example) and it clearly does not use three times the energy if it is open for three times as long. Until there are more data on this it is not possible to allow for 24/7 occupancy. Whilst it is acknowledged that this will be a problem for a few occupants, it will in the short term ensure that the system overall is more robust. For more information about the benchmarks see CIBSE TM46[15].

## 9.4    Floor area

Q1    Do I have to measure the floor area?

A1    Unless reliable floor area data are available from the client using the RICS approved method of measurement, then the floor area should be measured by the DEC assessor. If the area is incorrect and the rating is subsequently found to be wrong, the assessor will be held responsible. Given that the rating is based on consumption per unit of floor area, it is important that it is measured correctly.

Q2    I have been provided with some sketches of the layout of the building, which are marked as being to scale. Can I calculate the floor area from these?

A2    The CLG guidance is clear that the total useful floor area (TUFA) should be assessed using the RICS method of measurement. This does not currently allow for estimates from sketches or from any other forms of image derived from the internet.

The CLG guidance states that:

> ... the total useful floor area is defined as the total area of all enclosed spaces measured to the internal face of the external walls. Included are areas of sloping surfaces such as staircases, galleries, raked auditoria, and tiered terraces where the area taken is from the area on the plan. Excluded are areas that are not enclosed such as open floors, covered ways and balconies.

## 9.5          When DECs are required

**Q1**  I have read that the Directive actually requires certificates for all publicly accessed buildings. Is this correct?

**A1**  The preamble to the Directive, known as the Recitals, includes a suggestion that the scope should cover all buildings over 1000 m² frequently visited by the public. However, the Recitals are not considered by UK lawyers to form a part of the legally binding text of the Directive, and so the legal requirements are based on the wording of Article 7(3). This is explained in more detail in section 2.3.

**Q2**  Does a University hall of residence need a DEC?

**A2**  Yes, if it is larger than 1000 m². It is operated by a public body and accessible to students, who are members of the public.

**Q3**  Does a 300 m² nursery attached to 800 m² of unoccupied school need a DEC?

**A3**  It could be argued that if the nursery relies on the school for services, then the serviced area exceeds 1000 m² and a DEC is required. However, if the nursery is independent of the rest of the school, a DEC is not needed. This could equally apply to any combination of public buildings which are individually below the threshold, but together exceed it.

The CLG guidance is that where there is doubt about the need for a DEC, it should be obtained.

**Q4**  Is a DEC needed for a building with no heating?

**A4**  If energy is not 'used to condition the building' then the building does not fall within the scope of the Regulations and so does not need a DEC or an EPC.

**Q5**  Is a DEC needed for new-build?

**A5**  Where the building is owned by a public body or a public authority, then Regulation 18 applies. See section 9.1 for further details.

**Q6**  Is a DEC needed if there is no public access?

**A6**  If the building is not frequently visited by members of the public, then a DEC is not required. This has led to some debate about the requirement relating to prisons and other forms of detention centre and secure mental health facilities. In these cases there is almost always a right of access, albeit under strict controls, for visitors to those permanently resident at the facility. In the opinion of CIBSE these facilities should have a DEC.

**Q7**  Do halls of residence or care homes require a Display Energy Certificate (DEC)?

**A7**  Friends and relatives visit Halls of Residence, so a DEC will be required. Likewise a Local Authority care home where friends and relatives visit will require a DEC.

**Q8**  CLG Guidance talks about 'institutions providing public services to a large number of persons' and 'frequently visited by those persons'. What does this mean in practice?

**A8**  These are phrases from the text of Article 7 of the EPBD[8], and they are not defined elsewhere. The terminology is repeated in the Energy Performance of Buildings Regulations[1–7], which use the technique of 'copy out' to transpose the Directive into law in England and Wales. Whilst this ensures that the requirements of the European Directive are directly applied in England and Wales, it also directly transfers problems of interpretation where the Directive is unclear, as in this case. See section 2.3 for further guidance.

In the view of CLG, which CIBSE supports, if a building has a few visitors a week, then it requires a DEC. It has been suggested that where a large number visit just once or twice a year, for example to attend a military air show, or an 'Open Day' on a site normally closed to the public, this would not qualify. However, in the last resort this will have to be clarified in the courts.

## 9.6          DECs for campuses or sites

**Q1**  Under the transitional arrangements, a site had a single site-wide DEC. Will this be allowed when the next DEC is produced?

**A1**  CIBSE understands that this arrangement will not be extended, but that is a decision for CLG.

**Q2**  A site has no sub-metering. Can a site-wide DEC be produced, rather than individual default 'G's as they do not have separate building metering?

**A2**  Under the transitional arrangements a site wide DEC was permitted (see previous answer). However, in future each building over 1000 m² frequently visited by the public will require a DEC. This should be based on the metered energy use attributed to each building on a pro-rata basis. Defaults should not be used instead of metered data see section 4.2.

**Q3**  Does a school planned to be demolished need a DEC?

**A3**  There is an exemption from the requirement to produce an EPC for a building that is to be demolished, as long as certain conditions are met. If the school is still open then it will need a DEC, even if it is due to close and to be demolished.

**Q4**  Is there a website database holding the benchmarks for public buildings?

**A4**  Yes, these are held within the CIP. The data are also available in CIBSE TM46[15].

**Q5**  A school is being extended. How does the DEC assessor treat the extension and the energy used during construction?

**A5**  Energy used during construction that is part of the metered supply to the school is included. When the DEC is produced, it should be based on the floor area at the time of assessment. Additional benchmarks should also be added if appropriate. If the building has changed size (and possibly energy usage) then the DEC rating could change significantly from previous years. It is not possible to adjust or normalise the data so that the new DEC rating is in line with previous years DEC ratings.

Whilst one could undertake a series of meter readings and measurements during the process, ORCalc would not be able to handle this sophistication. It is understood that the DEC will be distorted, but if building work is underway then it is likely to be obvious to those most closely interested in the energy use of the school and due allowance made.

Q6 Does a DEC count towards CIBSE Low Carbon Consultant carbon returns?

A6 It does if you are asked to follow up on recommendations and turn some of the carbon counted into carbon saved. Otherwise you cannot include a DEC (or an EPC) in your return.

Q7 Can a DEC be renewed without a site visit?

A7 This is allowed as long as the assessor has a robust audit trail for the data used. A site visit is expected for an advisory report.

Q8 Can/should photographs be used by assessors as evidence?

A8 They can be very useful for audit purposes, to show evidence of the data used as the basis for a certificate or report.

## 9.7 Recommendations in the advisory report

Q1 The software tool has included two recommendations that are clearly not appropriate. It recommends installing double glazed windows and solar thermal panels, but the building is Grade II listed. What should I do?

A1 The tool allows the assessor to exercise proper professional judgement to exclude inappropriate recommendations and to add additional measures that the generic rules in the software tool have not identified as being suitable for the building in question.

This is an important aspect of the preparation of the advisory report. If the assessor just includes the recommendations offered by the tool without further consideration, then the client will not receive good value for the fees paid to the assessor. The assessor should review the recommendations generated by the software and consider whether they will be appropriate in the particular circumstances of the building being assessed. They should also consider whether there are additional measures that could be appropriate. The report writing tool also allows free text entries to be added. Again, this is an important feature for the assessor to use to make tailored recommendations to the occupier, perhaps in response to a specific behavioural aspect of the way the building is being used.

Q2 The advisory report requires me to make recommendations, but I do not have time within the fee to assess their detailed feasibility. What should I do?

A2 The standard wording of all advisory reports, which was prepared by CLG with the advice of government lawyers, includes a statement that the client should take advice before acting on any of the measures contained in the report. Therefore, as long as you have not included (or failed to exclude) anything in the report which is obviously not appropriate, your recommendations are offered to the client on that basis.

## 9.8 Occupied buildings

Q1 I have to prepare a DEC for a building which has undergone renovation in the last year, but the client has remained in occupation. What allowance can I make for the impact of the building works?

A1 There is no provision to adjust the metered energy readings to allow for any construction work. It is unlikely to be practical to meter the energy used by a contractor separately in such circumstances.

## 9.9 Estimated and absent meter readings

Q1 My client does not have utility bills with accurate meter readings — they have been estimated. However, the client's facilities manager read the meters at the start of the year and I have read them on my site visit. Can I use these readings?

A1 The CLG guidance suggests that the OR should be based on meter readings by utilities or their estimates. Unfortunately, it is widely known that at present many utility bills are estimated. Under the Energy Services Directive[16] the utility providers will be required to provide an accurate meter reading at least annually, which will help to address this problem. Where the energy assessor is presented with data that are not from the utility bill they may be willing to use these data if they can be sure that there is a robust audit trail to show that the data are reliable, although if they are in doubt then they should strictly apply the default rules and produce an Operational Rating of 200 and a 'G' grade on the DEC.

If incorrect data are applied in the first year, which understate the true energy use of the building, then there is a real risk that in year two the histograms will show a rise in energy use, and if this is large enough, it will also lead to a worse letter grade for the DEC the following year. As public awareness of DECs grows, then this could lead to some difficult questions in the local media, or even nationally if the building belongs to a high-profile public body.

Q2 My client is keen to have a DEC for the lowest possible cost. I have been asked to assume that there are no metered data and to issue a 'G' grade certificate with an OR of 200. Is this appropriate?

A2 A public sector body should make a little more effort to produce the data. Arguably it would have to respond to a freedom of information request for the data, so it should be willing to provide the data to the assessor. It is not in keeping with the spirit of the Directive or the Regulations in England and Wales to adopt this approach.

This approach may also potentially come back to haunt the client. If they are using considerably more energy than a typical building of the type, then a default OR of 200 may actually be unduly generous, and will result in a worsening of the reported performance in the following year. Again, this could have embarrassing repercussions for the public body occupying the building. It is a concern if a public body is either unwilling or unable to provide the data to enable them to comply with a piece of national legislation. The assessor might in this case wish to point out the risk of a problem the following year.

In year two there really should be no reason for not having accurate metered data for the start and finish of the year.

## 9.10          Penalties and enforcement

Q1      What are the penalties for not displaying a DEC?

A1      Failure to display a DEC is a civil offence, enforceable by the local Weights and Measures Authority, which is usually Trading Standards. They may act on complaints from the public or make random investigations. If they believe that you are affected by the Regulations, they can request you produce the relevant documents. You must provide this information within seven days of the request. They may take copies of any document you provide for inspection.

Failure to comply with either their request or the Regulations may result in the issue of a penalty charge notice. The penalty charge is £500 for failure to display a DEC at all times, in a prominent place clearly visible to the public, and £1000 for failure to possess a valid advisory report. In addition to the penalty charge, a DEC and a report must still be commissioned. Penalty charge notices can only be issued within six months of the date a DEC and advisory report was required.

If a public body can demonstrate that it has taken all reasonable steps to avoid breaching the Regulations, then the penalty charge notice may be withdrawn.

## 9.11          Site-based DECs

Q1      What is a 'site-based DEC'? And what are they for?

A1      For the first year of operation of Display Energy Certificates, CLG introduced transitional arrangements for 'site-based DECs' for sites with a number of buildings sharing a meter and requiring DECs. They were introduced by CLG 'to allow those organisations with a large number of affected buildings on one site to get ready in time for 1 October 2008.' The site-based DEC should cover those buildings on the site with a total useful floor area over 1000 m² which are also readily accessible to and frequently visited by the public. Buildings that are not frequently visited, such as research labs with restricted access do not require a DEC.

A site-based DEC is a single DEC covering all the qualifying buildings on a site. It has the same format, validity and other attributes as a building based DEC. If a building has its own meter, it

should have its own DEC and advisory report (AR). When the validity of the transitional DEC expires, the new DEC must be produced using standard OR methodology.

Q2      If I have a site-based advisory report, is that valid for seven years?

A2      Site-based advisory reports offer very little meaningful recommendations at a building level. CLG therefore 'strongly recommends' that site-based advisory reports are replaced by building specific reports for each individual qualifying building once the initial site-based DEC expires after 12 months. CLG guidance states that the site based AR 'must include a recommendation to get an AR for each qualifying building after the first year.' The AR for each qualifying building will then be valid for seven years.

Q3      Does a site-based Display Energy Certificate (DEC) mean that only one DEC and advisory report (AR) will be lodged for all the buildings on the site?

A3      There is only one lodgement, with a single DEC and AR to cover all the qualifying buildings on the site, which will have the site address. This is important for an assessor who is producing DECs for individual buildings in the coming year.

Q4      Where should a site-based Display Energy Certificate (DEC) be displayed?

A4      The site-based DEC should be displayed on every qualifying building on the site.

Q5      I have to produce a DEC for a school. There are three buildings, of 600 m², 720 m² and 1320 m² in area. The two larger buildings share a boiler house; the smaller building has separate heating. There is one electricity meter for the whole site. How should I calculate the energy used in the largest building?

A5      The gas consumption for the largest building will be the metered consumption multiplied by 720/1320 — this is the pro-rated area usage. It assumes that the two buildings use an equal amount of gas per unit floor area. Whilst this assumption may not be accurate, in the absence of sub-meters it is the recommended approach.

The electricity consumption is the metered electricity supplied multiplied by half, as the two smaller buildings are of equal area to the largest. Again, the same assumptions apply in this case. A recommendation from the assessor in the advisory report should be the installation of sub-meters to identify the actual use in each of the three buildings.

Only the largest building requires a DEC, as it is the only one over 1000 m².

## 9.12          Public display of DECs

Q1      Where should DECs be displayed?

A1      They must be placed in a prominent place clearly visible to the public within the publicly accessible space. However, some public bodies have taken a far more proactive approach by making their DECs accessible online.

The University of Leicester has a page devoted to the DECs for its campus, showing the grade of each building by colouring it on a map of the campus, with the colours following the colours of the bands on the DEC. This can be viewed at: http://www2.le.ac.uk/offices/estates/environment/energy/display-energy-certificates-decs

Reading Borough Council has published a list of the property reference numbers, which allows the certificates to be viewed on the Landmark® register of certificates. This can be viewed at: http://www.reading.gov.uk/Documents/community -living/DECs_RBC_Dec08.pdf

The Landmark® database is available at: https://www.ndepcregister.com/information.html

Torbay Council has gone even further, making its DECs available as PDFs on the council's website. These public bodies are fulfilling the intent of the EPBD, in making the energy use of their large public buildings transparent for all those who are interested: http://www.torbay.gov.uk/index/environment-planning/theenvironment/climatechange/localauthoritycarbonmanagement programme/displayenergycertificates.htm

# 10    Relevant websites

Further information may be obtained from the following websites:

— https://www.ndepcregister.com

This is the uniform resource locator (URL) for the register of energy performance certificates and display energy certificates for England and Wales.

— http://www.les-ter.org/page/home

LES-TER is an industry initiative led by the British Property Federation, in association with the Usable Buildings Trust, British Council for Offices and CIBSE, with financial support from the Carbon Trust. For further information see Section 8.

— www.cibsecertification.com

If you need an Energy Performance Certificate or Display Energy Certificate for your building then access the CIBSE Certification client area to find out how CIBSE accredited assessors can help you and the benefits of using an accredited CIBSE Low Carbon Energy Assessor. The site also gives details of training and accreditation for commercial energy assessors.

— http://www.ukreg-accreditation.org/index.php

Building Energy Calculation Software Approval Scheme: this website is the official website for the approval of software for use in connection with the Regulations. It outlines the procedure for the approval of third party software to:

(a)    demonstrate compliance with the carbon emission requirements of Part L of the Building Regulations for England and Wales

(b)    calculate Asset Ratings as part of preparing Energy Performance Certificates (EPCs) and Display Energy Certificates (DECs)

(c)    calculate Operational Ratings as part of preparing Display Energy Certificates (DECs).

— http://www.communities.gov.uk/planningand building/theenvironment/energyperformance

At the time of writing, this is the link to access CLG guidance on the implementation of the energy performance of buildings regulations.

— http://www.opsi.gov.uk/legislation/original

The website of the Office of Public Service Information, which provides downloads of the Regulations and amendments.

# References

1    The Energy Performance of Buildings (Certificates and Inspections) (England and Wales) Regulations 2007 Statutory Instruments 2007 No. 991 (London: The Stationery Office) (2007) (available at http://www.opsi.gov.uk/stat) (accessed 4/12/08)

2    The Energy Performance of Buildings (Certificates and Inspections) (England and Wales) (Amendment) Regulations 2007 Statutory Instruments 2007 No. 1669 (London: The Stationery Office) (2007) (available at http://www.opsi. gov.uk/stat) (accessed 4/12/08)

3    The Energy Performance of Buildings (Certificates and Inspections) (England and Wales) (Amendment No. 2) Regulations 2007 Statutory Instruments 2007 No. 3302 (London: The Stationery Office) (2007) (available at http://www.opsi.gov.uk/stat) (accessed 4/12/08)

4    The Energy Performance of Buildings (Certificates and Inspections) (England and Wales) (Amendment) Regulations 2008 Statutory Instruments 2008 No. 647 (London: The Stationery Office) (2008) (available at http://www.opsi.gov. uk/stat) (accessed 4/12/08)

5    The Energy Performance of Buildings (Certificates and Inspections) (England and Wales) (Amendment No. 2) Regulations 2008 Statutory Instruments 2008 No. 2363 (London: The Stationery Office) (2008) (available at http://www.opsi.gov.uk/stat) (accessed 4/12/08)

6    The Energy Performance of Buildings (Certificates and Inspections) Regulations (Northern Ireland) 2008 Statutory Rules of Northern Ireland 2008 No. 170 (London: The Stationery Office) (2008) (available at http://www.opsi.gov.uk/ legislation/northernireland/ni-srni) (accessed 4/12/08)

7    The Energy Performance of Buildings (Certificates and Inspections) (Amendment) Regulations (Northern Ireland) 2008 Statutory Rules of Northern Ireland 2008 No. 241 (London:    (The Stationery Office) (2008) (available at http://www.opsi.gov.uk/legislation/northernireland/ni-srni) (accessed 4/12/08)

8    Directive 2002/91/EC of the European Parliament and of the Council of 16 December 2002 on the energy performance of buildings ('The Energy Performance of Buildings Directive') Official J. of the European Communities L1/65 (4.1.2003) (Brussels: Commission for the European Communities) (2003) (available at http://ec.europa.eu/energy/demand/legislation/buildings_en.htm) (accessed 4/12/08)

9       *Getting ready for DECs* (London: Department for Communities and Local Government) (2008) (available at http://www.communities.gov.uk/publications/planningandbuilding/gettingreadyfordecs) (accessed 4/12/08)

10      *Improving the energy efficiency of our buildings: A guide to Display Energy Certificates and advisory reports for public buildings* (London: Department for Communities and Local Government) (2008) (available at http://www.communities.gov.uk/publications/planningandbuilding/improvingenergyefficiency) (accessed 4/12/08)

11      *Energy and carbon emissions regulations: A guide to implementation* (London: Chartered Institution of Building Services Engineers) (2008)

12      *Landlord's Energy Statement* (webpage) (London: British Property Federation) (available at http://www.les-ter.org/page/les) (accessed 4/12/08)

13      Freedom of Information Act 2000: Elizabeth II. Chapter 36 (London: The Stationery Office) (2000) (available at http://www.opsi.gov.uk/Acts/acts2000/ukpga_20000036_en_1) (accessed 4/12/08)

14      *Welcome to the Non-Domestic Energy Performance Certificate Register* (website) (Exeter: Landmark Information Group) (2008)

15      *Building energy benchmarks* CIBSE TM46 (London: Chartered Institution of Building Services Engineers) (2008)

16      Directive 2006/32/EC of the European Parliament and of the Council of 5 April 2006 on energy end-use efficiency and energy services and repealing Council Directive 93/76/EEC ('Energy Services Directive') *Official J. of the European Union* **L114/64** (27.4.2006) (Brussels: Commission for the European Communities) (available at http://ec.europa.eu/energy/efficiency/end-use_en.htm) (access 4/12/08)

17      *Code of Measuring Practice: A Guide for Property Professionals* 6th edn. (London: RICS Books) (2007)

18      The Building Regulations 2000 Statutory Instruments 2000 No 2531 as amended by The Building (Amendment) Regulations 2001 Statutory Instruments 2001 No. 3335 and The Building and Approved Inspectors (Amendment) Regulations 2006 Statutory Instruments 2006 No. 652) (London: The Stationery Office) (dates as indicated) (London: The Stationery Office) (2007) (available at http://www.opsi.gov.uk/stat.htm)

19      *Conservation of fuel and power in new buildings other than dwellings* Building Regulations 2000 Approved Document L2A (London: The Stationery Office) (2006) (available at http://www.planningportal.gov.uk/england/professionals/en/1115314231806.html) (accessed 4/12/08)

20      *Non-Domestic Heating, Cooling and Ventilation Compliance Guide* (London: NBS/Department of Communities and Local Government) (2006) (available at http://www.planningportal.gov.uk/uploads/br/BR_PDF_PTL_NONDOMHEAT.pdf) (accessed 4/12/08)

21      *Inspection of air conditioning systems* CIBSE TM44 (London: Chartered Institution of Building Services Engineers) (2007)

22      *Building log book toolkit* CIBSE TM31 (London: Chartered Institution of Building Services Engineers) (2006)

23      *Commissioning management* CIBSE Commissioning Code M (London: Chartered Institution of Building Services Engineers) (2003)

24      *Guide to building services for historic buildings* (London: Chartered Institution of Building Services Engineers) (2002)

25      *Minimum Requirements for Energy Assessors for Public Buildings (Display Energy Certificates)* (London: Department of Communities and Local Government) (2007) (available at www.communities.gov.uk/documents/planningandbuilding/pdf/publicbuildings.pdf) (accessed 4/12/08)

26      *Carbon Reduction Commitment* (website) (London: Department for Environment, Food and Rural Affairs) (2007) (http://www.defra.gov.uk/environment/climatechange/uk/business/crc) (accessed June 2009)